The Cray Valley Years

1946-2016

A history of Cray Valley Radio Society

Bob Treacher M0MCV

Published by Bob Treacher M0MCV

First published 2016

ISBN 9 781910 193297

Front cover: Kevin Williams M6CYB

Compilation, editing and typography: Bob Treacher M0MCV

Printed in Great Britain by 4edge Limited

Contents

Foreword

Introduction

Chapter 1:   In the beginning   1

Chapter 2:   The early years: the late 40s and 50s   7

Chapter 3:   Going from strength to strength: the 60s   13

Chapter 4:   From steady beginnings: the heady 70s   32

Chapter 5:   A decade of ups and downs   57

Chapter 6:   A decade in the doldrums: the 90s   83

Chapter 7:   Out of the doldrums...

turning on the activity tap: the 'noughties'   108

Chapter8:   On the home straight...

the recent past: 2010 to the 70th anniversary   162

Appendix I: Presidents and Chairmen   215

Appendix II: Membership 1946-2016   217

Foreword

In this 70th anniversary year of Cray Valley Radio Society, I am delighted that our President, Bob Treacher M0MCV, has written this comprehensive history of the Society, starting with its formation shortly after the war by eight radio amateurs.

With access to all but a handful of the Society's monthly newsletters, memories from Smudge G3GJW and pictures collected by Chris G0FDZ, Bob has been able to give a detailed account of people and events from the Society's formation up to the present day, including the 70th anniversary celebrations in October 2016.

Cray Valley has always been one of the foremost amateur radio societies in the country and had a strong record of special event stations long before the ground-breaking M2000A operation to see in the new millennium. Members have included three RSGB Presidents, G2MI, G3GVV and G3RZP, and many other notable amateurs but the Society has also taken care to try, as far as possible, to cater for the needs of all radio amateurs in its immediate area and beyond, and presently runs training courses at all three levels of the UK licensing system.

The idea of a written history of the Society was something of an accident. Chris G0FDZ had collected much archive material for his 'Project 70' endeavour and mentioned to Bob that a complementary booklet to guide people through the material would probably be a good idea. The 200 plus pages of this book are a direct consequence of the seed that Chris sowed!

The front cover of this book shows antennas from the 2O12L station put on by the Society for a period of 47 days for the London Olympic and Paralympic Games. This was surely the crowning achievement of the Society, but it drew on years of experience garnered from the earliest members when they created the Society in 1946. With membership levels currently stable at 100, I look forward to many more successful years as a Society which welcomes all radio amateurs in south-east London and beyond.

Dave Lawley, G4BUO
Chairman, Cray Valley Radio Society

Introduction

With the seed sown, the research began to record the highs and lows of Cray Valley Radio Society from its inception in 1946 to its celebratory anniversary party 70 years later. In conjunction with the publication of this written history, a large volume of memorabilia has been collected by Chris G0FDZ, and a memory stick is available to purchase separately.

I have had the privilege of being a member of this great amateur radio society since 1968, and its President since 1993. The 200 plus pages which follow provide as detailed a history as records have allowed. Full records from the late 1940s to the early 1960s are no longer available, but through recollection and archive material, I have been able to recount many facts and memories from that period.

The Society's history is told in more detail from 1962. Little did I know at the time how immensely valuable the copies of QUA obtained from Keith G3TAA before he moved away from the Cray Valley area would be. These have enabled me to provide a more detailed account of the Society through the 1960s and 1970s.

From steady beginnings, the Society enjoyed some heady times in the 1970s, with passionate activity and a membership of 149. The level of activity led to liability concerns and in 1977 the Society took the unprecedented step of becoming a company limited by guarantee. That status was partly responsible for the 1980s being a mixture of ups and downs, and the 1990s being a decade in the doldrums.

However, with members voting in 1999 to do away with limited company status activity returned to former levels and the Society quickly became well-known for its M2000A, GB50 and GB2000T special-special event stations, and in 2012 reached a new high by organising England's amateur radio celebration of the London Olympic and Paralympic Games.

Now, with a new shack, a healthy membership, successful training courses, and some enthusiastic and knowledgeable members, I sincerely hope the Society continues to build on the successes of the previous 70 years.

Bob Treacher M0MCV
President, Cray Valley Radio Society

Chapter 1: In the beginning

A limited meeting was called at the home of Les Allen G3MZ in Sidcup on 21 October 1946 to discuss the formation of a local transmitting Society to extend the friendships established over the air. The eight radio amateurs who attended the meeting were G2DS, G2HY, G2ZI, G2CXO, G3FC, G3MZ, G3ANK and G8PT. We are fortunate to have QSL cards for all of the Society's founder members.

G2DS proposed that Les G3MZ should chair the initial meeting. This was seconded by G8PT and passed unanimously. Geoff G2CXO provisionally agreed to undertake the Secretary and Treasurer duties. The aims discussed that night were to:

> *"(a) establish a discussion group and provide facilities for regular meetings of the local amateur fraternity for the purpose of promoting social contact, encouraging informal technical debate, and through a policy of mutual aid, to attempt to assist every one of us to get the most out of our hobby, and*
>
> (b) *encourage technical co-operation and mutual assistance on and off the air."*

These aims are as relevant today as they were then.

Membership was selective and later there was a waiting list for admission. It was proposed that membership be confined to licensed transmitting amateurs. At that time there were over 30 licensed amateurs in the district and finding a place to meet was seen to be the biggest problem. The meeting believed that if success was to be assured, membership had to be confined to those who had a genuine active interest in transmission.

The question of the admission of BRS members of the Radio Society of Great Britain (RSGB) was raised but it was considered that all such members had a keen interest in amateur activities and that to extend membership to include such people would be beneficial neither to themselves or to the Society as a whole. It was stated that under the licensing conditions at the time few obstacles were placed in the path of anyone desiring a licence and that this would be the only necessary qualification for membership.

Tentative proposals to be discussed in general discussion at a first meeting were:

"(a)   Discussion Groups. It is desired to break away from the formal lecture room atmosphere and to encourage free and informal discussion on all aspects of ham radio of a practical rather than theoretical nature.

(b)   Visiting Lecturers. G2DS and G3MZ are both confident that the services of well known personalities in the radio world may be secured for occasional lectures to Society members.

(c)   The XYL Problem. The conciliation of long suffering XYLs is an important consideration. At least one social function could be held annually at which any mention of radio would be strictly verboten.

(d)   Junk Sales. Comment unnecessary."

It was agreed that G2DS should take the chair at the first General Meeting held on Thursday 21 November 1946 at the Territorial Army Drill Hall, Halfway Street, Sidcup. All known local amateurs were informed that the Society was to be set up. And so the Cray Valley Transmitting Society came into existence. 34 members appeared on the Society's first membership list published in March 1947.

These few extracts from the 1947 'Cray Valley Directory' make interesting reading:-

G2CXO
G.Miles   Active on 7, 14, 28 Mc/S
"Cotswold"   Phone and CW
Mottingham Lane   Power: Phone 17 - 80 watts; CW 35 – 150 watts
Mottingham SE9   Xtals.   7019, 7091 and VFO
                               Calibrated RX. S. Meter

G2DHV
Geo.V.Haylock   Active 1.8, 3.5, 7, 14 Mc/S CW only
28, Longlands   Power: 25 watts (150 after 28.3.47)
Sidcup, Kent   Xtals. 1825, 3580, 7048, 7098

G3ANK
A.E.H.Swindon   14 Mc/S. CW only
135, Station Road   Power: 15 – 20watts (indoor ant.)
Sidcup, Kent   Xtals. 7075

G2ZI
R.C.Simmonds
164, Footscray Road
Eltham S.E.9.

Active 7, 14, 28 Mc/S. CW and Phone
Power: Usually 45 watts CW and Phone
Occasionally 130 watts CW
Xtals. 7023, 7127.

CVRS Founder Members' QSL cards

These cards were sourced in the main from Cray Valley Radio Society members, but grateful thanks are extended to the G4UZN Historic QSL Collection and Fraser Donachie for the G2DS and G3MZ cards.

From its inception, the Society made contact with its members through a newsletter which soon became a well-informed, well-produced monthly magazine. In June 1947 it appeared as "QRM" with a nine-page edition and a title cover featuring an agonised creature known affectionately as "QRM". It continued to appear for eight or nine years and after a short period of non-production, the newsletter re-appeared in 1956, less the title cover, ultimately to become "QUA" under the editorship of Brad Bradwin G3DNC in October 1964.

Early membership details
Apart from the eight founder members, archive records show that 10 other local radio amateurs were members of the Society in 1946. Their callsigns were: G2NK, G2WI, G2YZ, G3OW, G3PS, G3AEH, G3AQT, G3ATF, G6VV and G8WO. The first committee comprised of G2DS (Chairman), G2CXO, G3FC and G3MZ.

Based on archive material, 10 new members joined the Society in 1947 – G2AQB, G2FPP, G3AWD, G3BPE, G3CKL, G3CKW, G4DW, G5OH, G5SD and G6AG. In 1948, G3BTC, G3CLK, G3CSJ, G3CYX, G3EFQ and G6HK joined. Membership rose substantially in 1949 with G2BQ, G2YG, G2AXU, G2BQY (which is now the callsign used by the Trowbridge & District Amateur Radio Club), G2DHV, G3BR, G3CO, G3CKW, G3DBD, G3DEI, G3EHF, G3EIL, G3EIW, G3EMC, G3ENQ, G3ERP, G3FKC, G3FXY, G4CW, G4FZ, G6GK, G6HD, G6PY all becoming members.

Details of Society Presidents and Chairmen are shown at Appendix I, and a graph showing membership throughout the Society's 70 year history is at Appendix II.

The scene from 'The Bulletin'
The November 1946 Radio Society of Great Britain (RSGB) *'Bulletin'* showed a surprising world most had either forgotten or hardly knew. The editor was Arthur G2MI (not a member of the Society in those days) and the Advertising Manager was Horace Freeman. G2WS featured a regular "Month on Five" column; a band which did not exhibit the potential of the American six metre band. G5KT reviewed the first "Six months on the Top Band", which had been released the previous month extending from 1.7-2.0Mc/s. Arthur's "Month on the Air" column showed it was worth listening for Stateside stations on 11m and replying to them on 10m.

A construction article written by G5JU looked at a crystal controlled two valve transmitter for Top Band with self-contained PSU and impressive-looking iron-cored components by Woden. A 6V6 CO drove the EL37 PA and keying was provided but phone was only possible after modifications and the addition of an external modulator. An article by GM6MS gave full modifications for an R1155 which began by cutting off with tin snips, part of the DF section to accommodate an output transformer. By removing bands 4 and 5, 75-500kc with their large aerial coils, it became possible to fit an internal PSU.

It was reported that 15 members were present at the first meeting of a Woolwich/Plumstead based society held on 30 September 1946. Their members decided to hold future meetings on the third Monday of each month at St. Marks Hall, Plumstead Common. They beat Cray Valley off the mark, but did not manage to stay the course. In those days, 'London South' was in RSGB Region 13.

But back to 1946-47. The Society was a new, forceful, enthusiastic organisation as the first edition of "QRM" revealed. The Society entered its first National Field Day (NFD). 10 licensed members and one Short Wave Listener (SWL) helped to set up the station, but it was not as successful as had been hoped. That the Society had influential connections was shown by an article on Ionised Layers in that edition – based on a talk given to the Society by a Mr Bennington, who was regarded as the recognised expert on propagation at the time. Subsequent editions showed a ready co-operation by GEC Ltd to provide lecturers. G2DS contributed a dissertation – "Off the Pulley" on masts and troublesome fittings whilst "This Month's Station" featured G2YZ who served on the 1951/52 committee and was a member from 1946 through to 1957. He was called up at the beginning of WWII and posted to the RAF College at Cranwell to instruct young radio officers in the use of this *'new fangled'* radio equipment. At home, his station was situated on a roof top sun lounge, from where he listened to, and contacted, other radio amateurs. It is understood that he was the first amateur to contact The Flying Enterprise, a cargo ship which was sinking in a storm off the coast of Cornwall. He picked up a distress signal, alerted the Coastguard, and spoke to the captain, and the commander of one of the rescue tugs who had boarded the stricken ship. "QRM" appears to have bristled with contributions – a happy state of affairs for any editor – and set a standard of technical excellence that ran for many years. Although the "QRM" stencil had been kept with the Society's archives, no copies of the magazine were found during the search for memorabilia.

The Society's policy of excluding SWLs from membership caused some excited argument in those early days. There was an exclusiveness about the Society which might have been mistaken for aloofness by those who did not know the members, but it persisted. The Society remained outside the RSGB and adhered to an informality at meetings, not wishing to be tied down to lectures on fundamentals put on for the benefit of those seeking a licence. In this simple, exclusive manner, the Society built up a strong body of technical opinion that earned respect over the whole of Kent when there were moves to form an association of Kentish radio societies.

Stan G3JJC asked Dick G2ZI to reminisce at the time of the Society's 25th anniversary. He said his principal enjoyment came from the pleasant social atmosphere that marked the early meetings when technicalities and banter were exchanged in friendly rivalry. However, it seems things were not always calm because at one time some of the 'up-and-coming lads' apparently wanted to vote the committee out of office so things could be run differently. Wiser counsel prevailed, however.

At that time, Dick was celebrating his quarter century in amateur radio. His thoughts returned to 1921-22 and the thrill of his first transmissions when, as 5UO, he broadcast gramophone concerts on a frequency around 400m every Sunday from Ramsgate in Kent. Mains power was a considerable luxury in those days so he had to derive his HT from a Newton generator, belt driven from a stationary bicycle, furiously pedalled by G2HY. Accumulators for filament supply and dry batteries for grid bias completed the power arrangements for these transmissions, which attracted an enthusiastic weekly listener's report from a young Arthur Milne listening at Cliftonville, Kent and even comment from Hugh Redwood in the national *Daily Chronicle* newspaper. So you see, the foundations of wireless had a share in the founding of Cray Valley.

Also reminiscing was George G2DHV. He joined the Cray Valley Transmitting Society in 1949 when the Society met at the Station Hotel, Sidcup, but his first Cray Valley recollection was as an SWL hearing G8PT. George actually joined the Boys' Radio Society in 1930 and later, the Gravesend Amateur Radio Society and the Medway Amateur Radio Transmitting Society. As 2DHV, he became Hon. Secretary of the Sidcup and District Radio Society with G5SD as Chairman and Geoff G2CXO on the committee. Meetings were held at the Standard Telephones and Cables (STC) canteen at Footscray.

Chapter 2: The early years: the late 40s and 50s

Information relating to this period of the Society's history is relatively sparse compared to the other decades featured in this history. Even when Alan G4BWV compiled a record of members at the time of the 50th anniversary in 1996, he was unable to find precise details as no committee records were available for 1949 and 1950 and the period between 1953 and 1961. A dedicated search by Chris G0FDZ, resulted in some archive material being located, but much of the history of the Society between 1948 and 1961 remains unknown.

The first item of memorabilia from the early days of the Society is an item from the April 1948 issue of *Wireless World*, which gave basic details of the Society: that it accepted only licenced amateurs and met on the third Thursday of each month at the Adult Education Centre at Lamorbey Park, Sidcup. At that time the Secretary was Geoff G2CXO.

I am grateful to Alan G8BJG for trawling the pages of early editions of the RSGB *Bulletin* to provide the following details of Cray Valley's first contest activity and monthly meetings. The first recorded contest activity was an entry to National Field Day (NFD) in 1949. Published details showed 'Gravesend and Cray Valley' in 44th place: G6BQ/P and G2DS/P were listed as representing the societies. The Society consistently entered NFD between 1949 and 1955, although in some years entries were classed as 'Eltham and Sidcup' or 'Chislehurst and Sidcup'. G2NK/P, G3MZ/P, G6VV/P, G3ANK/P and G2YZ/P are listed as representing the Society. In the May 1951 *Bulletin*, G2BQY is shown as providing 160m slow Morse transmissions.

The March 1952 *Bulletin* refers to a first meeting of a *"newly formed Cray Valley RSGB Group"*. Details of the venue for this meeting were given as the Broadway Café at Southend Crescent, Eltham. Later issues in 1952 and 1953 refer to the Society as 'Eltham and Sidcup', meeting at Trinity Church Hall, Hurst Road, Sidcup. And during 1954 and 1955, references were to a 'Chislehurst and Sidcup' Society meeting at the *Seven Stars* at Footscray and then at the *United Services Club*, Sidcup, with Alan G3ANK listed as the contact. In 1959, references were to the

'Cray Valley Radio Club' meeting at the *Station Hotel*, Sidcup. By this time, Stan G3JJC was listed as 'Hon. Sec'.

Research has not identified why the Society used these different forms of identification.

Jeff VK6AJ (ex-G3JJX), who was never a member of the Society but attended meetings with Smudge G3GJW, passed on his reminiscences of the early 1950s when Alan G3ANK was stationed in Aden as a book keeper. Aden, of course, was a British interest in those days, keeping a watchful eye on Suez. With little to do in off-duty hours except drinking and socialising, Alan had the advantage of his amateur ticket and was soon on the air as VS9AS. Jeff was able to keep skeds with Alan, relaying messages from May, Alan's XYL, and gossip from Sidcup. On his return, Alan replied to the masses of QSL cards, and gave the Society a talk about his time in Aden at the Station Hotel in Sidcup.

To try to provide a fuller picture of Society activities in the early years, I am indebted to Smudge G3GJW for providing his recollections as a member of the Society in the 1950s. Some of these recollections dovetail well with the previous paragraphs.

Recollections can be somewhat hazy 66 years on, and there is a challenge to align events precisely with dates, especially when much written material had been lost during house moves, or sequestered for safety in dark places and forgotten. However, the following snapshots provided details which otherwise would have remained lost.

Smudge recalled his days training as a telegraphist. As a keen CW operator, thoughts of obtaining an amateur licence came naturally to him. With help from Don G3AQT, he took the Radio Amateurs' Examination (RAE) in May 1949, obtained the result in August and passed the statutory CW (Morse) 12wpm test at the South East London Post Office in September. He was licensed in late January 1950. The licence entitled him to transmit crystal controlled telegraphy with an input of 25 watts. This he did with a borrowed B2 spy set transmitter, run off the same power supply used for an American bomber receiver, BC348J, purchased from Bulls Radio at the end of the Northern Underground line for £19 12s 0d. After a year of 25 watt operation, his station was inspected and permission was granted for the use of amplitude modulation, 150 watts input and a variable frequency oscillator. Together with a dipole for 40 meters, the world of amateur radio became much more interesting especially as the peak of Solar Cycle 18 occurred around that time. He remained on CW for the first 20 years of his amateur career and used it to acquire many 'wallpaper' awards.

Smudge became a member of the Society in early February 1950. He preferred not to write specifically about individuals, but remarked that some members stood out as telegraphists extraordinaire; some built immaculate equipment. Clear recollections were of Alan G3ANK smoking a pipe and talking to a visiting friend at the same time as dealing with 35-40 wpm CW contact; and Les G3MZ, who worked for GEC in Greycoat Street, Westminster, turning out exhibition quality equipment, unequalled by anyone else in the Society.

The Society was composed of an interesting mix of amateur licence holders, many having served in the Forces who were known as having had 'an interesting time' during WWII. They had served around the world doing all sorts of things, about which they were reluctant to talk. Smudge recalled one member who had been an agent in France, but many had dealt with a bewildering variety of equipment and transmission modes, or had designed, built or serviced radio-related items. They had seen and worked in places which they would never have known about or visited had WWII not taken place. Spread about as they were, many never saw their school friends again.

The Society met at different venues in those early days. Smudge recalls meetings at Lamorbey Park Adult Education Centre in Sidcup, near Sidcup railway station, in 1949. Other venues included the *Seven Stars* in Foots Cray and the *Royal* in Mottingham.

Coronation special event station

Cray Valley was one of the first to operate two special callsigns at the same time. From 23 to 25 May 1953, members manned GB2ER and GB3ER in connection with the local Scouts' celebration of the Coronation of Queen Elizabeth II from a tent in the grounds of Eltham Palace. GB2ER used a Panda table-top transmitter on 20 metres with a ZL-Special antenna and a G.E.C. BRT 400B receiver. GB3ER, operating on 20, 40 and 80 metres, employed a Q-Max 40 watt transmitter, a Marconi CR 150 receiver, and a 'home-

A view of the operating position at GB3ER showing the Q-Max transmitter and Marconi CR150. G2WI is operating, with G3IGP at the receiver – from RSGB 'Bulletin' – July 1953

brewed' W8JK 20m reversible wire antenna sited to favour contacting Eltham in New Zealand, at the request of the Eltham UK Council. Unfortunately, New Zealand was not heard nor worked, but the special event logs quickly filled up with many American stations as well as enthusiastic UK and European ones. More than 100 stations in 20 countries were contacted despite a high local noise level. The July 1953 *'Bulletin'* reported that stations had some difficulty due to the unusual callsigns, which required frequent and lengthy explanations. The QSL cards for GB2ER and GB3ER became collectors' items.

National and International contests were entered, first on HF with 10w using specially built transmitters which were gradually replaced with commercial equipment running higher power, with single side band (SSB) on HF and frequency modulation (FM) on VHF and UHF making their appearances. The Society was rather 'nomadic' with many different portable sites chosen for their activities, which even today brought back memories for Smudge of lovely summer weather and/or torrential downpours!

Sites included the 'wet and nettle' field behind the Lamorbey meeting venue; Shooters Hill; the Royal Blackheath Golf Club alongside the 10th hole; Wrotham Hill, until a paging transmitter on the BBC mast ruined good CW reception; and at Telegraph Hill in Longfield in a torrential thunderstorm with sparking antenna feeders. Smudge remembers his father arrived unexpectedly as the station was closing, equipped with a bottle of brandy which was well received as the team were cold and wet!

Every site had its own story. At one, the team were visited literally in the middle of the night by an amateur who had travelled from Essex curious to find out why the station was receiving such good reports with their 100% legal 10 watts input. It seems the visiting amateur's team rigged an antenna across a lake and were using very much more power, and could not accept that the Cray Valley station was doing so well with legal low power.

Smudge remarked that the Lamorbey Park experience was *'interesting'*. He sold rocket sticks amongst other mass-produced wood products. With the aim of using rockets to elevate antennas into high trees, he asked a well-known fireworks manufacturer to make six without any stars, flashes or bangs. One was not required to be used on all occasions, but he tested one at about 03.30am during one portable operation. It vanished to a height of about 200 feet with a loud 'whoosh', which brought out the operators to inspect the tent by walking around it in very wet nettles. The story went into the folk lore of the Society!

Another site was on the ridge near Darenth, now part of a golf course. In short, there was scarcely a high location in the area which remained untried.

Social events were part of the Society's activities in the early years, with many different venues used, including Well Hall Pleasaunce at Eltham. On meeting evenings, every type of presentation and lecture was featured. This included his invention of the 'DIY Lecture' – colour slides of members' radio shacks, accompanied by their proud owners talking about their stations and activities. To be sure of members' full attention at the start of his presentation and after the tea break, Smudge used brief images of good looking ladies from the front covers of *Amateur Photographer*!

Cray Valley and RAYNET

When the Radio Amateur's Emergency Network (RAYNET) was founded in 1954, Smudge joined a weekly 80m net which included Alan G3ANK and Jeff G3JJX (now VK6AJ). Each used an individual crystal controlled frequency – his being 3.503 kHz. His first crystal was used with a B2 SOE government surplus transmitter, which he still owns today. The crystal cost 7/6d. Receiver selectivity was not too hot in those days. He used a BC348 from an American bomber, which he subsequently double super-hetted using 85kHz cans from an old command receiver.

The varying audio beats from each participant enabled instant recognition without the need to give callsigns – really good for 'break-in' operation. All three had message-handling experience in the Services and exchanged formal format RAYNET exercise messages very efficiently. Smudge and Jeff wrote the first RAYNET manual in 1954.

Cray Valley name change

As the Society was attracting local short wave listeners to its meetings, a decision was taken at the 1956 Annual General Meeting to accept SWLs as members: hitherto only licenced radio amateurs had been able eligible for membership of the Society. To reflect this change, the name was changed to 'Cray Valley Radio Society'. Accepting listener members was clearly a popular decision, as archive membership records show 13 SWLs joining in 1956 and in the following year, 24 SWLs were shown as members. The inclusion of SWLs as members increased membership of the Society from 16 in 1955 to 47 in 1957.

Archive records show the following licensed radio amateurs were members of the Society in 1957: G2HP, G2HY, G2NK, G2YZ, G2ZI, G2CXO, G2DZH, G3MZ, G3ANK,

G3BTC, G3CKL, G3EMC, G3FRB, G3FWI, G3HRC, G3HRO, G3ISX, G3JIT, G3JJC, G3KHJ, G3KIF, G3LCB and G3LWL.

Chapter 3: Going from strength to strength: the 60s

Understanding the history of Cray Valley Radio Society improved moving forward to 1962 as that is when the author's newsletter archive begins. In those days, the monthly magazine was edited by Stan G3JJC and was posted to members using a 2½d stamp.

Research revealed a great deal about how the Society developed during these years and the activities undertaken. The period appears to have been a successful one in which membership grew from 37 in 1960 to 73 in 1969.

Monthly meetings had been held at the *Station Hotel* in Sidcup since the late 1950s but because attendances had increased, members present at the 1962 Annual General Meeting voted to look to hire a room in which meetings could be held. The committee secured the use of *'a very good Society room'* at the Eltham Congregational Church Hall, 1 Court Road, Eltham SE9. The last meeting at the Station Hotel was on 29 September 1962. From July 1962, members kept in touch with Society affairs through a weekly 160m AM net, which was also followed by the Society's SWLs. The Society affiliated to the Radio Society of Great Britain as Cray Valley Radio Society in 1962, and G3RCV/A was activated for the first time from the 1st Swanley Scouts fete at Hextable on 21 July 1962. The first G3RCV logbook shows the initial contacts on 160m AM with G3DNC/M, G3OFM, G3OCC/M, G3BTC, G3PAZ, G2AQB, G3MCA, G3PJB, G3OZZ and G3JJC.

The autumn of 1962 was a busy time with a visit to the BBC Receiving and Measurement Station at Tatsfield and GB3RES operating at the 1st Royal Eltham Scout Headquarters in New Eltham for the 5th Jamboree on the Air (JOTA). 17 members took part in the JOTA activity. The Scouts provided a 45' mast made of lashed timbers and dural top section and found an anchorage for a halyard 50' up in a tree. Operation was on HF from a KW Vanguard transmitter, HRO receiver, KW trap-loaded dipole, and a 500 foot long wire. There was also a station on 2m. 165 QSOs were made. The activity was featured on the front page of the *'Kentish Independent'* and also the *'Eltham and Kentish Times'*.

The first official Cray Valley Christmas morning net was held in 1962, at 11.00am on 160m. It was very successful with 25 stations passing Christmas greetings.

1963

Due to local amateurs and SWLs hearing the Friday night 160m nets and good local publicity, membership swelled in 1963 to 52. 10 more licensed amateurs joined, together with seven SWLs. The Annual General Meeting reported a *'good year'*, although *'there was still room for improvement, particularly on the financial side.'* An attempt to raise subscriptions to 15 shillings was 'overthrown'; voting resulted in a £1 subscription, 7/6d for associates. Bill G3FBA was voted in as Chairman, with Stan G3JJC elected as Secretary. Dave G3MCA became newsletter editor.

The Society's first 'natter evening' took place on Wednesday 19 June 1963 at the Coldharbour Estate Recreation Hall, William Barefoot Drive, New Eltham SE9. 20 members attended and tea and biscuits were served free of charge after a 1 shilling entrance fee. An outing by 20 members was made to the ITV television transmitter at Beulah Hill, Norwood in July, and that month's talk was by Racal Electronics Ltd about the 'SSB abilities of the RA17'. This was followed in August by a 'factual report' by Dave G3MCA on the Post Office Radio Tower which was being built next to the Museum Telephone Exchange – did you know that the maximum deviation at the top (600') in high winds is 15 inches and that it weighed 13,000 tons?

Bill G2AQB earned the distinction of being the first to apply for the 'Worked Cray Valley' (WCV) certificate in August 1963, and Harold G3FS collected the first /M version in September 1963. With mid-monthly meeting attendances growing, G3RCV/A was aired from the Coldharbour Estate Recreation Hall. In September 1963 Steve G3KYV started an 'SWL Ladder' in the monthly newsletter. SWLs obtained points for hearing countries on 80, 40, 20, 15 and 10m, and additional points for obtaining a QSL card to confirm reception of the station. Also that month there was a talk about short wave listening which was attended by 40 members and friends.

The original 'Worked Cray Valley' award

The Society took part in the 6th JOTA weekend from the Kemnal's Own Baden Powell Scout Headquarters in Sidcup. Activity was on 160m through to 2m using three transmitters. This was a well-supported event organised by Steve G3KYV. It was supported by 18 members, who manned the stations for 41 hours. 250 contacts were made in 26 countries, including a large number of Jamboree stations. The newsletter waxed lyrical about the success of the weekend, especially being able to set up antennas without restriction. An entry was made to the *Short Wave Magazine* 160m 'Magazine Club Contest' in November. Using G3RCV/A from Rol G3NBT's QTH, Alan G3ANK, Stan G3AWD, Steve G3KYV, Harold G3FS and Brad G3DNC made 70 CW contacts. It was the first time in 10 years that the Society had put a contest station on the air. This was followed by the RSGB CW 160m Affiliated Contest (AFS) in February 1964.

1964

An interesting breakdown of Society membership by districts was published in the March 1964 newsletter. 19% of members lived in Sidcup; 18% in London SE9; 16% in Orpington and 15% in Bromley. Other members lived in London SE2, SE3, SE12 and SE18, Chislehurst, Welling and Dartford – even today that is traditional Cray Valley heartland.

Stan G3JJC was elected as Chairman at the 1964 Annual General Meeting, Steve G3KYV became Secretary and Alan G3ANK took over as Treasurer. Brad G3DNC took over as newsletter editor. In his Chairman's introduction in the May newsletter, Stan advised the membership that he was first introduced to the Cray Valley Transmitting Society in 1951 by G2ZI and G2HY and that it was a privilege to be Chairman *'to one of the livelier radio societies'*. Steve announced *'a new venture in the Newsletter'*. Because he admitted to *'not having a flare for the pen or as a scribe'* he hoped to include more articles from readers and announced the dropping of the detailed reports of meetings. It was hoped that the *'change in policy'* would give members an opportunity to be part of the newsletter's production. The May newsletter was the first to have an editorial and the first to have any sort of block header – albeit the usual typed details with a border using the '@' character.

The April 'natter nite' was reported as a *'great success'* with 27 members present. G3RCV was also on the air. Results of the 1963 'Ladder Contest' were published – the early fore-runner of the 'Annual Worked/Heard countries' table. There were quite simple rules – one point for each county contacted on 160m and 2m, one point for each country on the other bands, plus one point for every county or country confirmed. There were eight entries: Dave G3RGS won the contest with 138 points. Keith G3TAA and Ken G3TCC were announced as the latest members to attain full licensed status.

When was "QUA" born? Well, the May newsletter invited suggestions from members to give the newsletter a name, but more on the birth of the newsletter shortly.

By this time, the Society was deep in 'activity', with the Chairman's June newsletter referring to there being '...so much of it, we are becoming positively radio-active'. Thoughts turned to running raffles aimed at buying a transmitter for use in contests. Members were invited to bring items to be raffled, with a £30 target set. G3RCV/P was active in 2m Field Day from Heights View Poultry Farm making 68 contacts 'running 25 watts from a home-brew transmitter to a QQV 03/20, transistor modulator Class B, with an 840A receiver, converter, with power via a transistor converter from 12V car batteries'.

Much of the July newsletter was taken with a very full account of NFD: G3RCV made 177 contacts. Although a good time was had, the analysis showed an average of only 7.7 contacts per hour, or one contact every 7.8 minutes. It appears the performance was a topic at the following mid-monthly meeting. The July newsletter also reported that Stan G3JJC had received his 'Worked Cray Valley' Award.

A Directory of Members was circulated with the August 1964 newsletter. It showed which members were on the 'General' and 'Contest' committees, and listed 32 licensed members and 19 SWLs. It is interesting to see that current members Trevor M1TAD (ex-A4117) and Richard G8CDD were listed as SWLs. The newsletter also noted that Richard G3TFX and Phil G3TGQ had received their licences.

October 1964 saw the Society move its mid-monthly Wednesday meetings from the Coldharbour Estate Recreation Hall to "HMTS Kent", Ruxley Corner, Sidcup, Kent. There was no seating available for that first meeting as the chairs were locked away because no arrangement had been made to collect the keys to the room where they were stored!

'QUA' born

October 1964 also saw the birth of 'QUA'. Unfortunately, there is no mention of who suggested this name. However, 'I have news of...' seems a fitting title and one which still exists today. The heading was done by stencil and gave the newsletter a more professional feel. It was a bigger newsletter running to seven pages and spoke in glowing terms of the talk by Arthur G2MI as it had provided a refreshing reminder to the old-timers of the glories gone by and a stimulating reminder to all members of their responsibilities in amateur radio. From the rotary spark of North Foreland to the transistor and the International Telecommunications Conference, Arthur had seen it all within the span of his lifetime.

VHF Field Day, apparently in September in those days, was entered. G3RCV/P was active from a small field at Knockholt in the shadow of the masts of the Kent Police Radio. Operation was on 2 and 4m. Conditions were reported as *'difficult'*, which may have explained only 108 contacts on 2m and 69 on 4m. Nine members shared the operating. The October talk was 'My shack' by Smudge G3GJW. Smudge is our longest serving member, having joined the Society as an SWL in 1950.

First Cray Valley activity weekend

October also saw a Society 'activity weekend'. It was the first time such an activity had been arranged. G3RCV was active as were many members. The idea was mainly to publicise the WCV Award, but some stations thought it was a contest. The weekend was a huge success and Brad G3DNC considered it should be repeated as at least an annual event.

The NFD results had been released and gave members a boost as G3RCV was placed *'well above the halfway mark'*. Amendments to the Members' Directory showed eight new members joining since August. WCV Awards had been issued to G3BPE and G3JKY. Society members had also taken to entertaining overseas amateurs: G3MCA had welcomed W1VKR, SM5CKL and SM5AEX.

The December QUA had a really distinctive cover – an expertly drawn field day tent complete with operating station draped in holly, mistletoe and balloons! However, there was tragic news on turning the page with a report of the sudden passing of Dennis G3OYZ, who had been a leading light in the Society's contest exploits.

1965

It was announced in January 1965 that G3RCV would be operational from the Ruxley Corner meeting place on 160m from 2000z on each Friday commencing on 1 January. Colin G3SPJ and Keith G3TAA had several unsuccessful 'all-nighters' trying to contact transatlantic stations on 160m, as well as working to improve the 160m antenna. Once again, the 160m CW AFS contest was *'I fear not very good'*, with only 117 contacts made compared to several scores closer to the 200 mark. However, the photograph below has been dated using the Society's G3RCV log book as having been taken during the contest between 19 and 2300z on Saturday 23 January 1965.

Keith G3TAA and Colin G3SPJ operating G3RCV in the RSGB 160m AFS CW contest in February 1965

The venue was the Sea Scouts hut in Sandy Lane, Ruxley and the picture shows Colin G3SPJ operating and Keith G3TAA log keeping. Colin was 17 at the time and very keen on CW. The equipment used was Colin's home station, comprising a Collins TCS-6 AM / CW transmitter and his beloved HRO receiver, both of WW2 vintage. He was using a commercial single paddle electronic key which he bought from Smudge G3GJW at a Cray Valley surplus sale evening, but next to that is a WW2 army key that he still has to this day. Both pairs of headphones were also WW2 surplus and Colin recalls how very good they were. The blue cabinet to the right was made by Colin's late father Ken G3TCC and served as the kitchen range for family camping holidays with the cooker housed under the top lid and a ventilated larder underneath. As can be seen, Colin's family did not travel lightly in those days, but quite why the cabinet was at Ruxley he cannot recall!

The log shows that Colin worked the first 59 contacts of the AFS contest on the Saturday evening, with the remaining 58 contacts of the entry being shared by Keith G3TAA, Brad G3DNC and Colin on the Sunday evening.

18

Membership roll-call

At the start of 1965 there were 41 licensed members, for nostalgia reasons the list was as follows: G2HP, G2MI, G2ZI, G2AQB, G2CCH, G2DHV, G3FS, G3ANK, G3AWD, G3BTC, G3DNC, G3FBA, G3FWI, G3GAD, G3GJW, G3HRC, G3HRO, G3JJC, G3JYT, G3KYV, G3LWL, G3MCA, G3MCG, G3MIQ, G3NBT, G3OCC, G3OZZ, G3PXW, G3RGS, G3SDL, G3SOY, G3SPJ, G3TAA, G3TCC, G3TDU, G3TFE, G3TFX, G3TGQ, and, of course, G3RCV. Today, only G3GJW, G3RGS and G3SPJ remain members.

The April Annual General Meeting saw little change amongst the officers, but two short wave listeners were elected to the contest sub-committee. Following suggestions at the meeting, Ken G3TCC was co-opted as 'Social Secretary' to organise some 'social diversions', and it was announced that courses were to be organised at the Ruxley meeting place to help those intending to sit the Radio Amateurs' Examination. Arthur G2MI was elected as the Society's first President.

In May, a special event station was active from the Memorial Hospital, at Shooters Hill, Woolwich for the hospital's Annual Fete. *The Eltham and Kentish Times* singled out the Society for special mention referring to *"G3RCV, the famous Radio Cray Valley"*. Increased contest activity followed during the year, including a VHF Field Day entry from the grounds of the Memorial Hospital. Consensus was that 29th was *'not at all discouraging'* in view of troubles experienced with the 4m station.

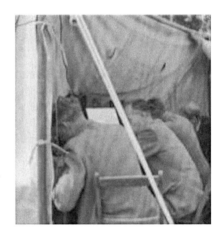

A 1960s Field Day event

Another activity weekend took place in October and Dave G3MCA took over the running of the 'Worked Cray Valley' Award. One of the best attended meetings of the year was Arthur G2MI's talk about the RSGB QSL Bureau, but perhaps the highlight was the revival of the Annual Dinner and Dance organised by Ken G3TCC and Smudge G3GJW. It was held at the *Bull's Head* in Chislehurst in November. Tickets were 25/-d and 78 guests attended.

1966

The year started in a flurry of activity with lots of work at the Ruxley Corner meeting place to set up a 'Radio Room'. Alan G3ANK had been particularly busy installing stable benching and making Improvements to the antennas. It was hoped that with *'...a fine shack, our own gear coming along and a good Society spirit we should be 'going places' this year'*.

The Annual General Meeting was something of a rarity when the previous committee was not re-elected en bloc. Brad G3DNC was elected as Chairman and the Society saw the first non-licensed member, Bill Davis BRS25995, elected to 'high office' as Secretary. This was also the first time Deryck (soon to become G3VLX) joined the committee as Treasurer. Other additions were Ken G3TCC as Vice-Chairman and Lyell G6HD, a very experienced and knowledgeable radio amateur. Alan G3ANK and Dave G3MCA stepped down. The contest committee was split into two – HF (G3ANK, G3GJW, and G3TAA) and VHF (G3AWD, G3OCC and G3TFE). These *'moderate'* changes were considered *'a good thing'*.

One of the heaviest items of Society expenditure had been the production and distribution of QUA. Certain concessions enjoyed the previous year had been lost because some the provision of free stationery had been lost. As a result, the Annual General Meeting agreed to increase the subscriptions of the senior members by 5/- a year and those members who wished to receive QUA also were required to pay a fee of 5/- a year towards the cost. The overall approximate cost of QUA was estimated to be 8/6d a year, per member. A Building Fund had been set up to reimburse members for out-of-pocket expenses incurred in the construction of equipment for the shack at Ruxley Corner. The necessary cash was to be raised by running frequent raffles. Interestingly, it was agreed to charge non-members 5/- a year if they wished to receive a copy of QUA.

With G3RCV and many Society members involved in VHF contest activity, the June QUA drew attention to a significant change in the new rules for the 2m portable contest. This marked the advent of exchanging 'QRA locators'. It is interesting to note that G3HRH, in announcing the change in the March 1965 issue of the RSGB 'Bulletin', used the term *'QRA locatormanship'*. The contest itself saw Cray Valley enter two teams: G3RCV was at Crowborough and G3POI at Ditchling Beacon. A shortage of battery power hindered G3RCV, but the value of sites well away from London was confirmed by G3POI achieving a good score and enjoying contacts with France, Holland, Germany and Belgium. G3RCV was placed 31st overall, with the 2m station 17th and the 4m station in 30th position. The team thought the improvement in the 2m score spoke well for the equipment and location but that it was obvious that improvements were required on the 70cms set-up if G3RCV was to work its way up the listings.

It is also worthy of note that QUA began to be a little more precise about timings of main meetings at the Eltham Congregational Church Hall. Doors opened at 7.30pm *'for the usual ragchew'*, business started at 8pm, with *'Lights out 10.30 sharp'*.

The first application from a YL (young lady) to join the Society was received in June 1966.

Planning for the 1967 VHF Field Day began early with the *'Bulletin'* announcing that the use of p.a. valves of greater anode dissipation than 13.5 watts would be forbidden. This ruled out use of the KT66, 6L6, 807 and 6146 valves, but allowed the use of the more modest 6BW6, 6CH6, 6V6, 5763 and TT11 valves.

A planned DXpedition to the Channel Isles for VHF NFD in September 1966 was cancelled for financial reasons; operation instead was from Knockholt in Kent, which proved to be a good site, helped by some Auroral propagation. Another successful activity weekend was held in October and a JOTA station was established for the 1st Royal Eltham Scouts, where the scouts made a good job of logging and writing QSL cards. The weekend was arranged by the Scouts, and one of the scout leaders was Derek Baker (who was to enjoy a long and successful association with the Society).

Questions were beginning to be asked about the success of the Ruxley Corner meeting venue. A committee meeting towards the end of the year felt that Ruxley as a QTH had not been all that it was imagined it might be. The main issue appeared to be that the venue was not a good radio location. Even with lots of hard work there was little to show by way of good results. A decision was taken to look for a change of venue for the 'natter nite' meetings.

1967

As a prelude to moving from Ruxley, surplus sales were held every Saturday and Sunday in February and March 1967 to sell up vast stocks of surplus Society equipment at modest prices. With the Ruxley location *'...not everyone's cup of tea...'* meetings there were discontinued. Instead, 'natter nites' were moved to the third Thursday of each month at All Saints Church Hall, Bercta Road, New Eltham SE9. The first meeting there was on 20 April 1967.

There were also mutterings about *'...the future well-being of our organisation...'* and a questionnaire was circulated to members to help the incoming committee shape future policy. However, less than half the membership returned the survey...did that mean that most members were happy with the status quo? Interesting comments were received suggesting that short wave listeners were being forgotten. Not the case was the committee's response noting that a listener representative, Derek Baker BRS28192, had been appointed to the committee to look after SWL matters and the forthcoming programme.

The new Bercta Road meeting venue attracted *'...a really good attendance of both licensed and unlicensed members'*. The reasoning was that it was more conveniently situated than Ruxley Corner.

NFD in the 60s
Looking at the year's external activities, the usual contests were entered with Chris G3VLT organising NFD – but only on 40-160m. Chris reported a *'...fairly successful...and reasonably smooth'* event. Colin G3SPJ recollects that the Society had a lot of very good CW operators in the mid-60s, all vying for the hot seat with NFD taken so seriously. This meant that only the top few were ever allowed to operate. New boys, like Colin, were thrown a few crumbs from time to time and allowed to log, whenever the operator's favoured logger took a comfort break. He believes that was how everyone cut their club contest teeth in those days!

Colin had better luck operating NFD during that period with G3BRK/P, the Aquila Radio Club (Aquila was the Electrical Quality Assurance Directorate of the Ministry of Defence based in Chislehurst, Kent which employed a number of local radio amateurs). Again there seemed to be no shortage of very good CW operators, with G3JKY being particularly prominent in that category. Colin recalls there was no shortage of excellent receivers either, with two mint Racal RA17s being signed out of G3HRCs communications laboratory especially for the task. The 10 Watt transmitters were home brew and beautifully made. Colin's contemporaries at that event were Keith G3TAA and Chris G3VLT, who were both Cray Valley members that also worked at Aquila. Alas, despite the luxury of excellent radio equipment, an army communications trailer and petrol generators that did not break down, success was scarce, but great fun was had by all. This account shows that today's 30 WPM part computerised 'Grand Prix' is a world away from the old NFD!

The following weekend in June saw an exhibition station at the Sidcup Military Tattoo and Trades Fair organised by Deryck G3VLX using the callsign GB2SMT. It was a '...*most successful day*' with the Mayor of Bexley taking particular interest in the station and the information displayed.

Contest results in 1967 were more pleasing; G3RCV was placed 16th in NFD and 14th in the 2m contest, and there was news of an enterprising young group of Society listeners who operated /P from Ashdown Forest in Sussex with a 20' whip antenna. They manned a 24 hour watch in July using CR100 and CR45 receivers and various antennas. They logged 230 stations, mainly on 160, 80 and 20m. All four listeners were to become well-known G8s within the Society.

Brad G3DNC had to relinquish the post of QUA Editor for work related reasons and Stan G3JJC took up the editorial reins again in August 1967. He had also advertised, successfully, for a skilled clerical assistant to fold QUA, insert it into envelopes, address, affix a stamp and post it. The Society gained its first Scottish member - David Douglas BRS26325 (later to become GM4ELV).

Cray Valley members 1967

L-R: Standing – G3KYV, G6HD, G3VLT, G3YGR, G3AWD, Derek Baker, G3JJC, unknown, G3WJK, G3TCC, G3NBT, G3GJW, unknown, unknown, G3ANK, G3GAD.
Seated – George Gregory, Phil Catchesides, Richard Buckley, G2MI, Paul Duthoit, Dave Howes

With thanks to *Practical Wireless*, Don G3XTT delved through the magazine's archives to find the September 1967 issue which featured the Society as the 17th in their regular 'Club Spot' series. The article provided an overview of the Society's formation, its early licensees, the twice monthly meeting arrangements, the weekly Top Band net and the 'Worked Cray Valley' award. Mention was made that the Society had shown little interest in contest operating until 1964, but since, both HF and VHF contests had been entered with varying degrees of success. In an attempt to co-ordinate the efforts of individual members, Chris G3VLT took on the job of Liaison Officer with the hope of improving results. The one big handicap was recorded as the lack of a permanent headquarters where equipment could be installed and aerials erected.

Experimental changes were made to the Society's 160m net in November 1967. Following a Thursday meeting, the net was moved to a Monday evening at 8.30pm, but it remained on a Friday evening on weeks where there was no meeting. The reason for the change was given that as so much talking was done at meetings, there was nothing left to talk about the following day!

Analysis of the earlier questionnaire was summarised in the November QUA. 29 responses were received. Although there had been criticism, the questionnaire responses suggested that things were pretty much OK: Thursdays were felt to be the best day of the week to meet; a majority were happy with the Society's contest activity; activity weekends were given the thumbs-up; and only two expressed dissatisfaction with QUA. There were almost equal numbers in favour of continuing the Annual Dinner and Dance, but there was less opposition to arranged outings. However, there was a call for more technical and instructional lectures, more effort to be spent on finding a permanent HQ, and trying to arrange classes for the Radio Amateurs' Exam. Perhaps the most disappointing aspect of the survey was further contention that SWLs were not made welcome at meetings.

As already mentioned, the first 'Society Directory' was published with the August 1964 QUA. In the December 1967 QUA it was noted that an offer had been received which would make it possible to produce a directory of paid-up members which included fuller details, including brief details of equipment and times and frequencies on which a member might normally be expected to be available if wanted. A form for this purpose was circulated.

21st anniversary celebrations

The Society celebrated its 21st anniversary in October 1967, but the celebration appears to have taken place in December as the January 1968 QUA provides details of the events. Quoting from the December 1968 QUA:

"The celebration seems to have…just grew and grew…resulting in the December meeting becoming …quite a success. It was good enough to earn a picture in the local press and a mini-report entertaining enough to hold the attention of the crowd who came along. G3VLX and G3TCC acting the part of the villain put on a bit of a pantomime to start off but soon got into his stride reminiscing over the experiences of the past year in which he showed that one's secrets are readily shared over the air. G3ANK looked farther back to the formation of the society and from a mass of original papers and records produced a very fine picture of the old days and the men who made up amateur radio in the late 1940's."

All those papers and records are, unfortunately, lost. It would indeed be fascinating to have access to them now to be able to reflect more thoroughly on the early days of the Society.

Cutting the 21st anniversary cake – L-R: G2CXO, G3ANK, G3JJC and G3VLX

A special cake was baked for the occasion by Joyce Buckley (Deryck's wife). Arthur G2MI officiated at the cake-cutting ceremony with a few well-chosen words of congratulation and reminded members to *'keep their records of today, which could well become tomorrow's history of amateur radio'*. Photographs were taken by the local press, there were ample cups of tea drunk and sandwiches munched, followed by more photographs, contacts with G3TAA's 4m equipment as G3RCV was aired. There were countless personal discussions and some tape recorded music before the celebration came to an end at 10.30pm. There was a feeling that the evening had been a real occasion, a meeting with a sense of history and achievement.

1968

An Extraordinary General Meeting (EGM) was called in January to approve a new constitution. After spending a great deal of time and care preparing the new set of rules, the committee were taken aback by the onslaught that apparently trivial items attracted. The EGM appears to have been no easy chat with a nice long tea break and one or two formal resolutions. It seems the members hardly got around to having a cup of tea as the session ran through three gruelling hours as clause after clause got its hammering. The committee were stung; was it vigilance, super caution, or merely a different slant to that envisaged by the committee? The constitution was eventually approved, and the hope expressed that the dangers and pitfalls so many members foresaw would never trouble future committees, and that comradeship and common sense would prevail with only the need for an occasional glance at the structure that had been devised to contain future conduct.

It is understood that a new membership directory was issued in either February or March, but no copy came to light during the search for memorabilia. The directory was produced by a short wave listener in return for the enjoyment he had in listening to Society members over the air. Unfortunately, there are no clues as to the donor.

Another activity weekend occurred in March, mainly to publicise the WCV Award – 43 licensed members were now available to those chasing the Award, including newly licensed members Bill G3XCJ and Fred G3XFG (who would soon have a big part to play in Cray Valley affairs). Previous comments about involving SWLs in Society matters had been heeded by the introduction of a short wave listeners' section in QUA.

The first 'professionally' produced QUA heading appeared in March 1968. It was prepared by Smudge G3GJW from Brad G3DNC's original hand drawn heading.

The road-mender's caravan

'Top secret negotiations' had been taking place to answer members' wishes for a permanent HQ. Fred G3XFG made available to the Society an old road-menders' caravan on his land at North Cray. After some sprucing up, it would be available for contests, trial runs and the like. As a permanent set-up it would enable the Society to get on the air with the minimum of delay and preparation; a most generous gesture by Fred. The caravan was just like the type that Fred Dibnah towed behind his steam tractor. It was painted dark green on the outside with a glazed wooden door and painted a faint pink on the inside with a single window on the side. It had large metal wheels which at some time must have had some type of rubber tyres and it had a large bench right across the far end with an over shelf for extra equipment and two large glass through insulators on the roof. There were a series of wooden steps up to the door from the ground. From the insulators, there was an end fed wire mainly used for 160m. The caravan was used only occasionally and one particular event that Chris G0FDZ recalled was during a 160m CW contest with Derek G3XMD operating and Alan G3ANK sitting behind Derek. Derek was sweating profusely. What he did not realise was that Alan had a heater directed towards the back of him for a joke! Chris was not sure if it was the heater or the CW (his first CW contest) that made him sweat so much! The caravan was not used as much as it should have been and eventually fell out of use, with Fred taking it back in the end thus ending our first real shack. Further contests from Fred's site, as you will read later, seemed to concentrate on using the nearby 'tack room' but this meant that no permanent station could be set up and left, although there was scope for some large wire antenna systems in the grounds.

There had been rumblings and differences of opinion about what the correct operation of G3RCV entailed. It appeared that some conditions of operation were not understood, for example, i) who may operate the station; ii) where may it be operated; and iii) whose is the responsibility for the operation of the station and the conduct of the operators. The situation was resolved when Deryck G3VLX put those questions to the Post Office and a most helpful reply was received.

Dave Howes acquired the callsign G8BKK and so became the first Cray Valley member to hold a G8 callsign.

Deryck G3VLX was elected Chairman at the April Annual General Meeting, but it turned out to be a short lived appointment. He was quick to go on the offensive in QUA. He reminded members that significant steps had been taken in 1967/68 to improve the running of the Society; the programme had been tailored to suit the majority of members and a start was made on a new QTH for G3RCV. He continued that if any member did not agree with the Society's policy or with the programme to tell him, but in doing so to offer an alternative suggestion or course of action. Firm words indeed!

What was billed as a *'Crisis in Cray Valley'* came about because the Annual General Meeting had left the Society in the awkward position of having elected a Chairman, but no Secretary. The first meeting of the new committee failed to find a solution so an Extra-Ordinary General Meeting had to be called in an attempt to find a Secretary. G3VLX tendered his regrets for the *'procedural error'*, and then his resignation! Lyell G6HD assumed the Chair. A vote of confidence in Deryck was supported by all, but he was not persuaded back to the Chair. However, he agreed that if nominated, he would take over the post of Secretary. He was elected without hesitation. But having secured a Secretary, a Chairman was required! Ken Wooff G3TCC was later elected. Constitutional *'crisis'* resolved!

More callsign successes were recorded in the June QUA as Derek became G3XMD, Harry, G3XOM and George became the Society's second G8 – G8BMT. Chris Whitmarsh also joined the Society.

NFD was again entered from the Memorial Hospital grounds at Shooters Hill. 248 contacts were made. The spade work was done by G3AWD, G3KYV, G3VLT, G3VLX and G3XMD, with the operating shared by G3ANK, G3DNC, G3RGS and G3VLT. G3TAA stood by as relief and Ted Bone (later to become G3XRX) did much of the logging. Catering was in the hands of George G8BMT and Chris (who was soon to be licensed as G3YGR). Encouraged by 248 contacts, the final placing was 23rd, but not as high a placing as obtained the previous year. In VHF NFD 27th place was achieved out of a total of 91 entries.

New shack at North Cray

The new G3RCV shack at G3XFG's QTH in North Cray was officially opened on 13 July 1968. A tape cutting opening ceremony was performed by Stan G3JJC. G3RCV was first aired at a garden fete run by the Scouts at Hextable in July 1962, but had subsequently been aired /A and /P from various locations. But for the first time, the permanent shack meant that 160m contacts made following the opening ceremony, using a Codar AT5 transmitter and an HRO receiver into a quarter wave antenna, could be made without adding the /A suffix.

The Society's second G3RCV log book began with contacts from the official shack opening. A number of members and others were contacted on Top Band AM. At the front of the logbook, Deryck G3VLX had this message for members using G3RCV:

"KINDLY NOTE: The GPO regulations must of course be observed in operating and completing the log. The log MUST BE COMPLETED AT THE TIME OF CONTACT and not written up later. Contest operation is NOT AN EXCEPTION TO THIS RULE. Please do not forget to include full details of rig and antenna, and QTH if away from 'home'. Without these it is impossible to QSL properly."

G3RCV was active in the 3rd 70MHz portable contest in July from Wrotham making 73 contacts. The Society also entered VHF NFD from the same site, making 148 contacts on 2m.

1969

The final year of this decade began with a talk by Fred G3SVK on his recollections from 160m exploits from the Shetland Islands. This talk was repeated to an almost entirely different audience in April 2016! The RSGB AFS CW contest was entered in January; interestingly, no contest numbers were shown in the log - suggesting the log was written up after the contest; clearly against the advice noted in the G3RCV log book! Later in the month, G3RCV took part in the CQ Worldwide 160m CW contest from the caravan at North Cray making 56 contacts. The February 1969 QUA was the first to be printed using the Society's new duplicator (who remembers Gestetner duplicators?), and Chris Whitmarsh became G8CIU.

The March QUA relived the story of a 23-man antenna party at Smudge G3GJW's QTH to raise a 50' telegraph pole weighing upwards of 30cwt.

New Cray Valley President elected

Stan G3JJC was elected as the Society's new President at the April 1969 Annual General Meeting, with Fred G3XFG elected as Chairman. Derek G3XMD filled the Treasurer's seat, while Ted G3XRX, Don G3KGM and Ian Lever (who was to become G8CPJ) filled the committee posts. Ian was also commissioned to build, on behalf of the Society, a 2m transmitter for use at the caravan. The transmitter would be in modular form with a 15 watt PA, 15 watt modulator, switched crystal frequency change, multiple PSU, and the capability to be used on other bands. An

FET converter was to be provided, feeding into an HRO which Alan G3ANK had donated for Society use at the caravan.

Due to decorating at All Saints Church Hall, natter nites in May and June 1969 were moved to the Eltham Congregational Church Hall, but the Society did not return to the former venue, deciding to hold both monthly meetings at Court Road, Eltham.

Peter Vella G3WVP joined the Society in April 1969 (he was to have a big presence in Cray Valley activities for many years) and John Thomas was licensed as G3YDW. May 1969 saw a field day station set up at the 9th Royal Eltham Scouts' May Fair at the GLC Playing Fields on Rochester Way, Eltham by a group of the Society's younger members – G3YGR, G8BKK, G8CIU and Richard Buckley (who would be licensed as G8CTT). July saw GB2SMT on the air again from the Sidcup Military Tattoo and Trade Fair. Equipment was a KW2000A, courtesy of KW Electronics, into a KW Trap Dipole. DX contacts were plentiful, including some Japanese stations.

Mindful of the need to protect the Society against the possibility of a serious claim through damage (for example if a mast at an exhibition station should fall down), July 1969 saw the Society take out Public Liability insurance cover of £100,000.

G3RCV/P was active in NFD in June. Operating from the grounds of the Memorial Hospital at Shooters Hill, contacts were made on 20, 40 and 80m using dipoles.

From an idea by SWL Chris Pearson A-5935, an SWL prefix table started in September 1969. One Bob Treacher, who had joined the Society the previous year headed the first table, but it was discontinued before the end of the year due to poor support. My amateur band listening had started shortly before I was invited to a Cray Valley meeting by Ted G3XRX – we were both reading *Short Wave* Magazine travelling home by train from work in London. Visits to the shacks of local amateurs Eddie G3LLT and Stan G3AWD soon followed as I became more involved with Society matters.

The 1969 VHF NFD was something of a milestone when G3RCV appeared for the first time on 70cms. 15 members took part – a mixture of G3s, G8s and SWLs – and stations were run on 4m (G3TAA/P), 2m (G3YGR/P) and 70cms (G3RCV/P). Best DX was to GM (Scotland) on 4m, but results were disappointing – outside the top 20 again.

As the decade came to a close, GW3HUM became the first non-English station to qualify for the 'Worked Cray Valley' award, and G3RCV was active in the 160m Magazine Club Contest from the caravan at North Cray using a Codar AT-5 (10 watts), BRT 400 receiver and an 80' end fed antenna.

Chapter 4: From steady beginnings: the heady 70s

The 70s decade saw Cray Valley Radio Society thrive and become one of the best supported amateur radio societies in the British Isles. Under the chairmanship of Deryck Buckley G3VLX, Derek Baker G3XMD, Martin Tripp G3YWO, Bernard Harrad G8LDV and myself, membership peaked at an all-time high of 149 members, and attendance at meetings necessitated a move to a larger meeting venue. The decade also saw the Society begin to take part in more international contests, with some success, collecting some worthwhile achievement certificates in the process.

1970

Membership had risen from 73 in 1969 to 79 at the beginning of the year, and the Society's growing SWL contingent managed to convince the committee to reinstate the SWL table – the first one showing four entries. Also in the January QUA was an article written by me on short wave listening which attracted a degree of interest and led to Chris Pearson claiming a CHC (Certificate Hunters' Society) award for hearing 50 DXCC entities on 160m.

G3RCV operated from the caravan at North Cray for the RSGB AFS CW contest in January. Using a Codar AT-5 and half-wave long wire, 119 contacts were made.

It is interesting today to note the joy expressed following the first surplus sale of the year which made a £30 profit! Contest success came early, with Alan G3ANK and Chris G3VLT winning the 80m Field Day, picking up the Houston Fergus trophy. Martin G3ZAY joined the Society in January 1970 (and was soon to be a leading light in our international contest activity). Fred G3XFG was re-elected Chairman at the April Annual General Meeting, with Ken G3TCC Vice-Chairman, Don G3KGM Secretary and Derek G3XMD Treasurer. I was elected to the committee, and it appears to have been the first year that the Society elected an auditor (Alan G3ANK). Despite the ever mounting cost-of-living, annual subscriptions were maintained at £1 for 'corporate' member and 10/- for 'Associate' and 'Country' members.

A new team was fielded in NFD, again from the site at Shooters Hill. The contest station had a number of visitors - patients and staff from the adjacent hospital! In the contest G3ANK, G3XRX, G3TAA and G3YGR made 262 contacts. The following week a special event station was activated for the Sidcup Trade Fair at Sidcup Place, Sidcup. The event was organised by Deryck G3VLX and Derek G3XMD, who were to become the mainstay of external Society activities through the 70s. The callsign used was GB3STF. Using a Sommerkamp FL200 transmitter and Yaesu FR100-B receiver into G3ANK's Mosley TA33 Junior yagi and a 40m trap dipole,

162 contacts were made in 45 DXCC countries by seven operators. The RSGB provided a backdrop and table drapes, and the stand was popular amongst visitors, including the Mayor and Mayoress of Bexley.

G3ZAY, G3VLX, BRS32525 and G3XMD at GB3STF

Deryck and Derek set up another special event station the following weekend at Bexley Borough Council's Donkey Derby at Crook Log, Bexleyheath, using a Sommerkamp line and an inverted-V antenna. The Society launched its own short wave listener contest in this year.

VHF NFD took place from Terry's Lodge Farm, high up on the North Downs at Wrotham, organised by Chris G3YGR and Chris G8CIU. Three stations (2m, 4m and 70cms) made a total of 278 contacts and were placed a creditable 36th out of 119 entries. GB3RES was on the air for JOTA from the Royal Eltham Scouts District HQ in New Eltham SE9. The event was co-ordinated by Derek G3XMD and was covered by the *Eltham and Kentish Times*. Scouts camped in the grounds overnight and, using a KW2000B transceiver, 103 contacts were made using a trapped dipole antenna in an inverted-V configuration.

1971

With greater publicity, the 'Worked Cray Valley' award had suddenly attracted much more interest, with European stations seeking out members. A second Cray Valley SWL contest took place in March: 23 logs were received. The winner was Len Randall (who was to become G4ACQ). The SWL table became the 'Heard/Worked' table this year. The first table had eight entries, led by Martin G3ZAY (the popularity of this table was to grow and grow). The final table had 13 entries, with SWL Tony Whittaker (who became G4GLB, and is now EA3GLB) overtaking Martin G3ZAY to take first place.

From May, in the interests of economy, the publication date of QUA was changed so that copies could be collected by members at the mid-monthly meeting to save postage. Until this point 'For Sale' advertisements had appeared in QUA free of charge. However, due to a perceived loss of commission had the items been sold at a 'Surplus Sale', a charge of 5p per advertisement was introduced.

The Society obtained its own G8 callsign – G8FCV.

Fred G3XFG stood down at the Annual General Meeting and Deryck G3VLX was elected as Chairman as the Society entered its 25th year, with Ian G8CPJ as Vice-Chairman. Peter G3WVP was Secretary. Arthur G2MI became a Vice President. QUA remarked that *'the committee had a good feel about it'*, with full licence holders (G3VLX, G3WVP, G3XMD and G3MCA); G8s (G8CPJ and G8CIU); and SWLs (me) represented. Subscription rates remained the same, largely due to the excellent support members afforded to surplus sales. Dave G3MCA regretfully resigned from committee later in the year. His place was taken by Martin G3YWO (another to have a long and successful role in Society business).

NFD in 1971 was again from the Memorial Hospital grounds at Shooters Hill; Ted G3XRX operated G3RCV/P and made 83 contacts. GB3KBC was put on in July from the grounds of Lord Maserene & Ferrard's Castle at Chilham, Kent to coincide with the Annual General Meeting of the Kent Association of Boys Clubs. Nick G3YQG, Richard G3YJW, Peter G3WVP and I took part; a KW204 transmitter and a Trio JR500S receiver into 15, 20 and 80m dipoles were used. The year was dominated by a full programme of contest and other activity. The Society participated in VHF NFD, the Worked All Europe contest, organised an activity weekend, another SWL contest, held a successful Dinner and Dance, entered the CQWW SSB contest, and followed these activities by celebrating the 25th anniversary of the Society.

VHF NFD is worthy of fuller mention as Chris G8CIU organised it and oversaw what was believed to be a successful contest with many high scoring contacts made. Again, the site was at Terry's Lodge Farm. 19 members were involved, with G3TAA/P on 4m, G3YGR/P on 2m and G3RCV/P on 70cms. However, the results were somewhat disappointing (52nd out of 129 entries), which led to Chris volunteering to issue 'an operating guide' ready for the following year's event.

The Society was informed by G3ZOJ that he was interested in re-forming the Woolwich Radio Society which was at one time, quite a thriving organisation, affiliated to the RSGB and one of the signatories to a petition presented to the Postmaster General on 29 December 1921.

The September activity weekend was run, for the first time, as a contest. It was reasonably successful with eight entries and three checklogs. Stan G3JJC achieved the best score, followed by Norman G3ZCV and Derek G3XMD. 52 entries were received for the Society's 3rd SWL contest – amazing support!

A special eight page anniversary issue of the October/November QUA was prepared to celebrate the 25th anniversary of the Society: four pages looked back, while four gave news of the Society's recent and forthcoming activity. The front cover reproduced the "QRM" cover from early Cray Valley days. In October, GB3WS represented the 1st Welling Scouts, organised by G3WVP, and G3RCV/A operated on behalf of the 1st Royal Eltham Scouts at New Eltham. At GB3WS, the high spot was their ability to stay on one frequency on 80m for eight hours and 20 minutes!

G3RCV was active again at the end of the month, this time from Martin G3ZAY's parents' QTH in Petts Wood for the SSB leg of the CQ Worldwide contest. Peter G3WVP loaned the Society his caravan for the weekend and it was parked on the drive. The weekend was, unfortunately, not trouble free as Martin's FTdx560 failed and the station caused a great deal of BCI to one of the neighbours, who opened their lounge windows wide to let us hear the S9 'Donald Duck' noises coming through their Hi-Fi speakers! 11 members took part in this useful first attempt at a major international contest. 561 contacts were made, but the team knew there was much to learn to be able to be competitive in contests of this kind.

The KW line and G3ZAY operating G3RCV in the CQ Worldwide contest

Also in October, a joint dinner and dance was held with the North Kent Radio Society. Colin G3VFD proposed a toast to *'Our two Societies'* and congratulated Cray Valley on its 25th anniversary (the North Kent Society was formed in 1947). Tim Hughes G3GVV, RSGB President elect, was the Guest of Honour. The event was a financial success and the same venue was booked for the following year.

The anniversary was celebrated at the Society's main November meeting in a light-hearted manner, with reminiscences backed by relics, photographs and archives, together with tea and cake. G2ZI, G3MZ and G3ANK were present to turn the pages of the book that was unknown to most attendees. They demonstrated too, that the world of amateur radio in those early days, including the illustration of newsletters, was very much a DIY affair, but at least radio amateurs had a ready supply of components and adaptable units from a generously stocked government surplus market.

1972

A new QUA heading, designed by Derek G3XMD, was unveiled in the January/February issue. QUA also changed to being printed on A4 paper, rather than foolscap. News was given that negotiations were in hand for leasing a new Society room to be used for natter nights with the chance of being able to house a Society station. Funds, however, needed to be supplemented to obtain all the equipment needed for a station that would hold its own. The purchase of equipment was authorised by those at the Society's January meeting and a special 'bring and buy' sale was arranged for the following month to help raise funds: £25 was raised. An HF net was established on Sunday afternoons to generate more HF activity for the WCV Award. QUA also noted the appearance of my SWL column in Radcom, which came about as a result of a talk given to the Society by the editor of Radcom; the column was to appear for over 30 years.

Derek G3XMD was elected Chairman at the Annual General Meeting. He updated members on progress of the Society shack project, and the work being done to improve activity and interest for the Society's listeners, including provision of an RAE course by Martin G3ZAY and Richard G3YJW out of University term-time, and Martin G3YWO at other times. Plaques, donated by Chas G5GH, were awarded to Martin G3ZAY as 'Top transmitter' and to Tony Whittaker as 'Top SWL' in 1971. The introduction of plaques to the leading amateur and SWL increased participation in the annual table: 18 at the year end, won by Cris Henderson (to become G8KNW and G(M)4FAM).

L-R: G4AYO, G3XRX and BRS32525 at NFD

NFD was entered with the North Kent Radio Society; a horse box at G3XFG's QTH providing the operating shack! Much of the effort Ted G3XRX put into the entry was justified by the results achieved. In July, G3RCV/A was aired from Chilham Castle, Kent, for the Annual General Meeting of the Kent Association of Boys Clubs. In August, a large contingent of members attended the RSGB Mobile Rally at Woburn. Three RSGB stations were active – G3VHF, GB3RS and GB2VHF. Martin G3ZAY came away with top prize in the raffle – a VHF FM/AM receiver.

VHF NFD was again entered from the site at Terry's Lodge Farm, Wrotham. It was a well-supported activity with 27 members involved. JOTA again saw a two-pronged effort – GB3WS and GB3RES. The station for the Royal Eltham Scouts benefitted from the use of my new TA33Jnr tri-bander, bringing in DX contacts with Tristan da Cunha, Ethiopia and the Dominican Republic. 94 contacts were made, including 32 JOTA stations. Operators were G3VLX, G3XMD, G3JJC, G3TAA and G3ANK. The Society also put on a multi-single entry in the SSB leg of the CQ Worldwide contest from G3XFG's QTH in North Cray using an FTdx560 into a TA33Jnr yagi, plus inverted-V dipoles for 40, 80 and 160m.

G3RCV entered the 160m Magazine Club Contest from Keith G3TAA's QTH in New Eltham. The 4th SWL contest was really well supported with 67 logs received, including entries from Rhodesia, Swaziland and Japan. The Society was filling a gap in the SWL activity calendar and listeners from all over the world were appreciative of the Society's efforts on behalf of SWLs. Four Cray Valley SWLs were well-placed in the 1972 *Radio Communication* SWL table, with Cris Henderson A7460 topping the table.

QUA often offered rather tame jokes, but this one made me smile –

First old lady:	*"Isn't it windy!"*
Second old lady:	*"No, I think it's Thursday"*
Third old lady:	*"So am I, let's have a cup of tea".*

1973

A list of WCV Award holders was published in January. Some familiar, but some unfamiliar, callsigns were listed. The full list, at that time, was:

G2AQB, G3MCA, G3FS/M, G3LCB, G3SDL, G3SPJ, G3JJC, G3SZX, G3HSE, G3PGI, G3JKY, G3BPE, G3SVK, G3UJE, G3FVG, G3VLT, G3XFZ, G3ZAY, G8CIU, GW3HUM, WA2DWE/2 and K4AUL, with G3YWO soon to qualify for certificate No. 20(G).

This listing marked the introduction of an advanced class WCV award. 45 points (three for a member and a specified number of special event stations, and one for Society stations and special event callsigns above the specified number) would enable a gold seal endorsement to be claimed. To help publicise the award, stickers with the award rules were produced for members to affix to outgoing QSL cards. G3VLX was to be the first claimant.

Martin G3YWO began another RAE instruction course in January in readiness for the May RAE. The January activity weekend was quite well supported, winners were: HF G3ZAY; VHF G8FUR (now G4DCV). Karen Ramsay (now G8JNZ) won the VHF SWL section.

A first HF contest success

Members' approval was obtained in February to purchase a yagi and rotator for use in HF contests. A TA33Jnr and AR22R rotator were purchased and G3RCV was active in the ARRL DX SSB and CQWPX SSB contests. 547 contacts were made in the ARRL DX SSB contest from G3XFG's QTH using a KW Atlanta and an FTdx560. It is interesting to note that the contest exchange given was "59 600"! In those days, of course, logging was done with pen and paper, so one of the operators always had the job of transcribing the contacts to the official contest log sheets after the contest! CQWPX SSB at the end of March saw the first recorded antenna erection 'incident'. When trying to erect a 55' mast and TA33Jnr yagi with the aid of a gin pole, the mast buckled in the middle. This type of incident has been largely uncommon in the Society's history, but the term "Cray-Valleyed" is a term still referred to today! 18 members helped in some way or other, and

745 contacts were made. This was enough to bring the Society home in 1st place in England – our first HF contest win (8th in Europe and 22nd in the World). Also in March, G8FCV/A was given a rare airing by Richard G8CTT in the 144/432MHz Open contest. Operators were G8CIU, G8CTT, G8EZM and G8FUR.

The CQ WPX SSB contest certificate

The Annual General Meeting saw Derek G3XMD elected for a second term, and I was elected as Vice-Chairman. New to the committee were Owen BRS33364 (to become G8HYH and now G4DFI) and Richard G8CTT. Alan G3ANK was installed as the Society's second Vice President.

The June meeting provided a record attendance of 51 when Ian G3PRR and his XYL, representing Western Electronics, put on an impressive show of sleek Japanese hardware. The following month's QUA leader noted the fine job the committee were doing to attract new members and getting Cray Valley better known. By the year end, membership was to rise to 90; 17 higher than at the end of the previous year. A natter nite even attracted 38 members!

Activity continued into the autumn with entries into NFD, VHF NFD, the Worked All Europe DX Phone contest, the 5th SWL contest, the Annual Dinner and Dance, the Scandinavian Activity Phone contest, an activity weekend, Jamboree-on-the-Air (GB3WS), and the CQWW DX contests (both SSB and CW). Phew!

CQWW was not quite as successful as CQWPX. Operating from G3XFG's QTH, G3RCV amassed a best-ever contact total of 1,586 and a claimed score of 1.2M points. Martin G3ZAY made 500 QSOs, Derek G3XMD over 400, with 200+ QSOs from G8HYH, G3VLX and G3ZRJ. However, G3RCV was only placed 5th in England, but was 21st in Europe from 124 entries, and 42nd in the World from 212 entries.

The final 'Heard/Worked' table of the year saw 23 entries. SWLs Dave Churchill, Cris Henderson, yours truly and Owen Cross took the first four places. Top licensed member was G3ZAY, followed by G6HD, G3VLX and G4BAL. Two YL SWL members also appeared in the table.

Some of the Society's more established members were given some Christmas present ideas in the festive QUA. Having known each quite well, these suggestions really were spot on!

G3ANK: *Modulator for his Top Band rig*
G3JJC: *Automatic Morse reader*
G3XMD: *A DX aerial*
G6HD: *'Variable Frequency Antenna" (for testing – can be disposed of as an unwanted present if found unsuitable).*

1974

Together with the increased, and more successful G3RCV contesting activity, and the success of the SWL contests, the Society's profile was being raised both at home and around the world. By the year end, the Society would be able to boast 109 members.

Congratulations were due early in the year when four SWLs – Charles Abrathat, John Balsdon, Bob Mersh and Roger Smith - were successful RAE candidates. John and Owen G8HYH passed their Morse test and became G4DEY and G4DFI.

The contesting theme continued into 1974 with entries into both the ARRL DX CW and SSB contests from our contest QTH at G3XFG's QTH. Using a KW Atlanta into a TA33Jnr yagi, 40m and 80m delta loops and a 160m Inverted-V at 40', 804 stateside contacts were made, but G3RCV was not to win that one! The team were surprised to find a large rabbit hutch in the shack. Some went to explore but all that could be seen was…hay. Empty, we thought. Suddenly, in the middle of the contest a rustling noise was heard from the vicinity of the hutch. Upon further investigation we saw a guinea pig amongst the hay. It was quickly given the name 'Russell' (Rustle – get it?). But suddenly there were two Russell's, and then three, and four, and five! The contest crew became really fond of them, but it was pointed out that Russell must have been a 'Russellina'! An entry into the CQWPX SSB contest followed in March, with 1,038 QSOs and 3rd place in England.

Martin G3YWO was elected as Chairman at the Annual General Meeting. Owen G4DFI became Treasurer, and Norman G3ZCV and Cris Henderson joined the committee. Arthur G2MI, on behalf of the Society and representing the RSGB, presented the *CQ Magazine* certificate for our 1st place in the 1973 WPX contest. A vote of thanks was given to Derek G3XMD for his *'splendid efforts'* as Chairman during the previous two years. It was noted that he had been *"…a never failing source of energy and a tower of strength from which the Society had benefitted."* Arthur also presented the annual 'Top Transmitter' plaque to Martin G3ZAY. The first two Advanced WCV Awards were presented to Deryck G3VLX and Peter G3WVP, while newly designed 'Heard Cray Valley' awards were presented to Robbie Robinson, Cris Henderson, Karen Ramsey, Dave Churchill and myself. The meeting voted to increase subscriptions to £1.50 for Corporate members and 75p for Associate and Country members.

The April/May QUA looked back with some pride at the Society's historic first HF contest success. It was the day that the contesters had been waiting for since the entry into the Worked All Europe contest in 1971. It seems that Adrian G3RUV, one of the successful at the time Exeter Contest Group G3WYX, said it had taken them three years *'to get somewhere'*. So it was good to note that G3RCV had taken the same amount of time to win its first international contest.

G3RCV took part in the All Asia DX contest in June from G3XFG's QTH. Using an FTdx560 and an FL2000 linear amplifier into a TA33Jnr yagi and wire antennas for 40 and 80m, 70 contacts were made. This was sufficient to secure another 1st place in England certificate, but this time we were also 1st in Western Europe.

The Society featured in an excellent article about the year's activity in *The Eltham and Kentish Times* in July. Another activity weekend took place in July, and in September G3RCV/P was active in SSB Field Day. With Martin G3YWO's RAE classes flourishing, Alan G3ANK began running Morse sessions later in the year at Nick G8EZM's QTH. Weekly 80m CW practice sessions were also arranged for Saturdays and Sundays.

The Society hired a coach to take a party of 28 to the RSGB Woburn Rally, but 30 boarded the coach. Little thought was given to the two strangers until it came to collecting the £1.25 fare. They asked where the coach was going. 'Woburn', was the reply, but they wanted to go to Eastbourne! The couple enjoyed the trip very much and expressed their surprise at finding such genial company.

Historic special event station

VHF NFD was entered from a site 8k NE of Sevenoaks. G3RCV/P was used on 23 and 70cm. G3RCV took part in the Worked All Europe SSB contest in September making 446 contacts. GB3WS was used in JOTA in October, and a Society team took on and beat the Reigate Amateur Transmitting Society in a radio quiz evening, but on 19 October 1974 Cray Valley struck a historical note by activating a special event station to commemorate the 50th anniversary of the first contact

between England and New Zealand. GB2SZ was operational from Arthur G2MI's QTH using an FTdx560 into a TA33Jnr yagi at 60' to celebrate the 1924 contact made by Cecil Goyder, a London schoolboy at Mill Hill, and Frank Bell Z4AA in Palmerston, New Zealand. The station was activated on behalf of the RSGB.

A great deal of assistance was given by G8PO, G6XN, G6DW, VK3AD ex-G6TM and ZL4BX.

Members of the Otago branch of the New Zealand Amateur Amateur Transmitting Society assisted Frank Bell in putting ZL4AA on the air and contacts were made on 20 and 80m. ZL3AD, who was a friend of Cecil Goyder, was also contacted. The Society's thanks went to Paul WB2OZW who personally contacted Cecil Goyder, who was living close by in Princetown, New Jersey. The following message was received from Cecil:

"Warmest greetings and best wishes to Frank Bell on the 50th anniversary of our first QSO, England-New Zealand. I am sure you join with me in greeting and appreciation of the Amateur Fraternity who have thus recalled this memorable occasion. Signed Cecil Goyder ex-G2SZ."

This was a truly noteworthy occasion commemorating one of the milestones of amateur radio history.

Concluding a busy month, G3RCV was active in the CQWW SSB contest from G3XFG's QTH. 1401 contacts were made, not as prolific as in 1973, but again the magic 1,000 contact barrier was achieved; G3RCV was placed 3rd in England. In December, G3RCV entered the ARRL 10m contest, again from G3XFG's QTH. The final 1974 'Heard/Worked' table showed 17 entries, with 10 entries to the VHF equivalent.

1975

The year began with a combined entry into the RSGB's AFS CW contest with the KW Radio Society. Peter G3RZP provided all the equipment except the mast and antenna. G4AUU/P was used and gained a best CW contest result – 5th place. Ted G3XRX, Fred G3XFG and Cris Henderson joined Peter to man the station. QUA reported that I made the tea (and opened the ring-pull cans)! The KW Society callsign was used again for a joint effort in NFD in June. It was the first year the Society had entered with such seriousness for some time. Peter G3RZP did much of the organising, but the Cray Valley presence made the difference with Paul G3SXE and G3XRX joining Pete for the operating. Logging and check-logging was carried out by a number of Cray Valley's G8 and SWL members. The 'G3VLX 40m Challenge Trophy' was introduced by Deryck to be awarded to the member with the best yearly DXCC country count on 40m.

The first weekends in February and March saw entries into the two-leg ARRL SSB DX contest. Operation was from the Victory Social Club in Hayes as our usual contest QTH was unavailable. Arthur G2MI had kindly arranged for the premises to be available, together with his TA33Snr yagi. 40/80m wire antennas were also installed. For the second weekend of the contest we returned using a TA33Jnr at 60' and a new 80m Delta Loop antenna – the brainchild of G8ON, Peter G3RZP's father. A further 600+ contacts were made.

This year saw the 'Cray Valley Contest Group' heading to France to compete in the CQWPX SSB contest at the end of March using Deryck G3VLX's call F0RV from Paul F2YT's QTH at Arras in northern France. The team comprised G3VLX, G3XMD, G3ZAY and G3ZEN. Cris Henderson and I were also members of the group. It was a cold, snowy weekend with the station set up in Paul's large garage. 1,231 contacts were made. 1st place in France was achieved, together with a place in the European Top 10. A 50+ attendance heard all about the trip at a Society meeting later in the year.

I was honoured to be elected Chairman of the Society at the Annual General Meeting; the first, and so far only, SWL to hold the position. Peter G3WVP was elected as Vice-Chairman. In fact, three SWLs were voted on to committee, with Cris becoming Treasurer, and Denise BRS34902 becoming the first YL committee member and taking over the SWL column in QUA. G2MI presented the annual awards, including Advanced Cray Valley Awards to G3XRX and G3WVP.

The May meeting attracted an audience of 60 for a lecture on the subject of 'Radio Paging'. The long awaited new 'Cray Valley Award' became available in June. The timing was appropriate as all copies of the earlier certificate had been exhausted. New G3RCV QSL cards were printed, to be issued exclusively for contest contacts; I was the Society's QSL Manager at that time.

The contest calendar was full for the remainder of the year with entries into seven HF/VHF contests between April and September, including airings for G8FCV/P on Shooters Hill for the 2m portable event at which a new 14-element Parabeam was used, and from Bob G8JNZ's QTH for the RSGB Jubilee contest. It is worth noting a best ever (at that time) NFD placing of 16th for G3RCV. Another G3RCV 1st place certificate was achieved for the entry into the All Asia DX contest from Arthur G2MI's QTH in Hayes. VHF NFD was a particular success this year with a 4th place finish – more details below. A Society activity weekend took the form of a four hour 160m and 2m contest. It proved to be one of the most successful with 47 logs received; G3RCV was active from Martin G3YWO's QTH, with Mike G4AYO making the CW contacts. Paul G4BXT was 1st on 160m, with G4BWG 1st on 2m.

Although not a Society event, G8CIU, G8GGP, Cris and Denise DXpeditioned to the Channel Islands in June, operating on 2m and 70cms from Jersey, Guernsey, Sark and Alderney – the second DXpedition mounted by Cray Valley members during 1975. In August, G3RCV entered the Worked All Europe CW contest from Hayes making 412 contacts. In September, 381 contacts were made in the Worked All Europe SSB contest.

Christina G4CMM and Nigel G4AXA wed and became Cray Valley's first husband and wife amateur transmitting team. 17 members made the trip to the Reigate Amateur Transmitting Society in July for a long-awaited return quiz. The Cray Valley team of Peter G3RZP, Deryck G3VLX and Tim G8GGP were victorious, thus completing a notable double. Several listener members obtained their G8 licences – Bernard G8LDV, Cris G8KNW, Kevin G8KXB and Mark G8LAJ.

A coach was again arranged to take members to the Woburn Rally in August; 40 members made the trip and *Radio Communication* published a photograph of members at the rally. Three special event stations were arranged – for Eltham College, the Victory Social Society's fete in Hayes (the site for most HF contest activity at this time), and GB3HVS from the Venture Scout HQ at Hither Green for the JOTA weekend; 21 members helped man the JOTA HF and 2m stations.

VHF and UHF contest successes

VHF NFD this year saw a 14 person team head to QRA locator ZK21F (the Isle of Purbeck, 3k NW of Swanage in Dorset). The site at 654' ASL had a take-off over the water on three sides. A total of 478 contacts were made over the four bands. With good weather, a good site and good tropospheric conditions, the 70cms station (G8AYN/P) was placed 3rd, and G3RCV/P operating on 23cm was placed 5th Overall, 4th place was achieved showing the worth of travelling out of the Cray Valley area so contacts with stations in the London area provided better point values.

An entry into the RSGB UHF contest in October was even more successful. A group of members operated from Butser Hill, just south of Petersfield in Surrey. The team included G8CIU, G3TAA, G8AYN, G8IWX, G8CTT, G8GGP and G8JNZ. The contest was noticeable for some good tropospheric conditions on 70cms into Germany. The team eagerly awaited the results. They were not disappointed as Cray Valley swept the board! G8AYN/P was overall winner, plus 1st on 70cms, and G3RCV/P was 1st on 23 and 13cms. This was a splendid result that reflected the hard work and bounding enthusiasm demonstrated by the VHF/UHF group at that time.

October saw the Society's best effort up to that point in the CQWW SSB contest. Over 1,800 contacts were made, but this was only good enough for 4th place in England. In those days G4DAA, G3UBR, G4ANT and G3WYX were the most successful English multi-operator contests teams.

Towards the end of the year, a Cray Valley technical library was established by Peter G3WVP. It included technical reference books and articles which were available to members on loan. A Buffet/Dance at *The Woodman* at Blackfen was particularly successful. QUA reported the music being 6dB down on the previous year, which enabled the talkative types to make themselves heard, but there were also reports of *'music rolling and liquid refreshment flowing'* which appears to have led some members launching into good old English folk dances like the *Gay Gordons, Knees-up Mother Brown* and the *Hokey-Cokey!*

At the year end, members entertained themselves at the *Park Tavern* in Eltham with great enthusiasm. QUA reported it as *'...the biggest turnout we have ever had'*. As well as a table loaded with food, members were entertained with folk songs by G3VLX and G4DFI, while G3TAA played a mean piano and G3XMD played popular male voice choir songs! The Christmas morning net was very successful, with 28 stations calling in, together with 'landline' contributions from a number of VHF members. The final 'Heard/Worked' tables showed 21 HF entries and 11 VHF entries.

The Society reached the end of 1975 on a rising tide of success. Could 1976 get any better?

1976

The Society's year started early, on 1 January. 40 plus members attended a New Year's Day meeting. The attendance book showed that the number attending natter nites was growing considerably, although not equalling the average 50 attendance at main lecture nights. A get-together with the Channel Contest Group (G4DAA) was arranged at *The Felbridge Arms* in East Grinstead, Sussex later in the month. As you could imagine talk was mainly contests!

The Society entered a new-style 80m AFS contest as G3RCV/A from Tony G3ZRJ's QTH and were placed 7th of 51 entries. A Cray Valley activity week was arranged for the final week of January; this included short contests on 160m, 2m and 70cms. Maurice G4BAL operated G3RCV/A in the 160m contest. There were 13 entries. Also in January, it was announced that Morse classes were being run by Alan G3ANK.

At the end of January, the G3RCV logs were scrutinised to calculate DXCC: 195 entities had been contacted. This was further broken down as follows: 28MHz – 89; 21MHz – 152; 14MHz – 175; 7MHz – 76; 3.5MHz – 68; 1.8MHz – 12. Total band countries totalled 572.

The Annual General Meeting was an unusually long affair. It noted that membership had reached a staggering 163 (although this figure is not supported by archive material). But with that came issues of overcrowding at meetings: we were increasingly being told by the Church Hall caretaker that we were in contravention of the fire regulations. Members also debated concerns about the continuing financial stability of the Society and about their personal welfare in view of the possibility of incidents occurring at an external event or at a contest. These issues were high on the committee's agenda throughout the year. I was voted in for a second term as Chairman. Bernard G8LDV, Glyn G8KKI and SWL Roger Smith (to become G4FSZ) joined the committee. Due to the ever expanding membership, Owen G4DFI became 'Membership Secretary'.

G3RCV again exceeded 1,000 contacts in the CQ WPX SSB contest in March from the Victory Social Club grounds in Hayes. In May, a special event station was set up for *'The County Patrol Challenge'* at The Fort in Dulwich, while in June G4DFI, G8JNZ and G8LDV put on a station for the Belvedere Youth Association in Belvedere. And in June a third special event station was established for the Addey and Stanhope School Parent/Teacher's Association at the Whitefoot Lane Playing Fields, Downham.

G3RCV/P was active from a site at Farningham for NFD. G3RZP, G3PSV, G3XRX and G4BXT made 512 contacts. VHF NFD was again from the Isle of Purbeck in Dorset, but with no repeat success. A successful first Construction Contest saw 18 entries judged by G3FRB, G3VLH and G4DMO from the Reigate Amateur Transmitting Society. They donated 'The Reigate Cup', which was awarded to John G3HRO for his 80/160m solid state phasing rig.

In September, Cray Valley joined forces with the Medway Amateur Radio Society for the 2m Open contest from a disused World War 2 anti-aircraft gun site near Chattenden Barracks, just outside Hoo on the Isle of Grain. The main talking point from this was the fire lit for warmth at around midnight in an old fireplace. With the timbers damp, smoke filled the shack as the chimney was blocked. The problem was partially solved, but the night crew of G8ECI, G8FUF, G8GGP and G8JNZ completed the operating stint smelling like a bonfire!

30th Anniversary celebrations

October saw full and varied celebrations for the Society's 30th anniversary. Special event callsign GB3OCV was obtained and was used throughout the month from Arthur G2MI's QTH. Even by Cray Valley standards, the first meeting of the month was a phenomenal success with a 70+ audience for Roger Smith's 'Meet the Members' slide show. The mid-monthly meeting was also well-attended with a buffet, wine, and celebratory cake made by Miles Buckley (Deryck G3VLX's son) and topped with a model of G2MI's tower and yagi. This provided a worthy foil to the ceremonial cutting by two founder members, Dick G2ZI and Alan G3ANK. The evening was further marked by the appearance of Lynne, Norman G3ZCV's daughter, newly licensed as G4FNC. Lynne was the Society's second ever licensed female member. A special Cray Valley Award was also available, and two well-supported activity weekends were arranged to make it easier to collect the 30 points required to claim the award. The activity weekend contests were a success, with 44 logs received. A new G3RCV QSL card was designed to celebrate the 30th anniversary. It included the callsigns and receiving station numbers of all members as at July 1976. The *Eltham* Times reported the activities, and QUA noted the total attendance for the two October meetings was over 160. The month ended with an entry into the CQ Worldwide SSB contest.

Continuing the celebratory theme, over 100 Cray Valley and North Kent members enjoyed a thoroughly enjoyable time at the *Tudor Barn*, Eltham at a November buffet/dance – all except one that is; Norman G3ZCV's wife had the misfortune to slip on the polished floor and fracture her ankle.

As a result of the earlier Annual General Meeting debate over the Society's financial stability and members' personal welfare, an Extraordinary General Meeting was called for the mid-November meeting. Fred G3XFG reported on the Society's insurance policy against claims for damages. Further debate ensued and a vote taken on whether the Society should become a Limited Company under Guarantee. The vote recommended acceptance of the proposal to limit the Society's liabilities in the hope that the move would lead to increased activity and an influx of new members. Steps were taken to form a Limited Company and membership forms were sent to all members.

The pre-Christmas meeting was again held at the *Park Tavern* in Eltham – Deryck G3VLX hosted a quiz and sang folk songs and a selection from *Gilbert and* Sullivan accompanied by Steve G3YPK on guitar, and on occasion by the audience, with other musical effects provided by Derek G3XMD.

By the year end, the 'Heard/Worked' QUA table listed 32 entries, and archive material showed that membership had risen to 132. It is not known why this figure differs from the 163 referred to at the Annual General Meeting.

1977

Members believed the rules of the 'Worked Cray Valley' award had become too easy, so the committee were tasked with investigating ways of making it more attractive and less easy to obtain. New rules were announced with contacts counting from 1 January 1970 instead of 1 January 1962. Bernard G8LDV took over as librarian for the Cray Valley reference library, which listed 78 titles. With attendance numbers rocketing – there were 90 plus at the January meeting – steps were being taken to secure larger premises to provide more breathing space at meetings.

I was elected for a third term of office as Chairman at the Annual General Meeting. Bernard G8LDV became Vice-Chairman. Paul G3SXE and Peter G4FUG joined the committee. Due to the hiatus of Limited Company status, the North West Kent Contest Group was formed by 10 members keen to get back to participating in serious contest activity. The group was formed for the specific purpose of entering the CQ WPX contest from a site at Hinxhill, near Ashford, Kent. 1,319 contacts earned the group 2nd place in England.

1977 saw Queen Elizabeth II celebrate her silver jubilee. The Society organised its own 'Queen's Jubilee' contest in June, with separate 10m and 160m sections on different days – Arthur G2MI aired G3RCV on 28MHz and Deryck G3VLX used it on 1.8MHz. 37 entries were received and certificates were awarded to the various section winners. In July, G4FAM and G8JNZ operated /A from the Eltham College summer fete using the college's own equipment. Members rallied round to talk to the public and an exhibition of Morse keys, coupled to audio oscillators, displayed by Ted G3XRX, was popular with the students. Also in July, a team made its way to Dorset for VHF NFD, but no report appeared in QUA, and Roy G3JHI organised a special event station at the Lambeth Country Show at Brockwell Park.

Cray Valley Radio Society Limited formed

Fred G3XFG handled all the complex legal technicalities enabling Cray Valley Radio Society Limited to be formed on 7 July 1977. The Memorandum of Association and Articles of Association were signed by the members of the Board. The Memorandum of Association, whilst imposing certain administrative disciplines, also incorporated a very wide range of outputs (21) to encompass the Society's organising abilities and energies. Establishing the Limited Company therefore limited the liability of members. QUA noted that this was '...a remarkable status for a radio Society'. It was achieved after a great deal of work by the committee – now the Board of Directors.

The CVRS Limited Company seal

The first meeting in August gave a glimpse of the future when a talk by the North West Kent Repeater Group about 'The 70cm Repeater' was held at the *Christchurch Centre* in Eltham High Street. The Centre, a roomy and well-appointed venue, would become Cray Valley's new meeting place from 5 January 1978. Also in August, Tony G3ZRJ and Roger G4FSZ were the operators at a special event station at the Victory Social Club for their summer fete. G3RCV made an entry into the CQWW SSB contest from the same site in October, but with a smaller team – Peter G4FUG, Keith G3TAA, Derek

50

G3XMD, Arthur G2MI and Roger G4FSZ were the operators. After the success of the previous Construction Contest, entries for the 1977 event were disappointingly low.

Moving to the end of the year, the Society featured in an article about amateur radio in the *Kentish Times* in November. Changes to the Board occurred in December as Martin G3YWO moved away from the Cray Valley area due to business commitments. There was also a resignation. Thanks were expressed to Martin for the service he had afforded the Society. He had been Chairman, Secretary and Treasurer, and had also provided successful RAE tuition. Martin was also the first Company Secretary of the Limited Company and had shouldered a considerable burden translating extensive expertise in company law for the Society. Peter G4FUG joined the Board. Archive records do not show who resigned.

Another successful annual social function was held at the *Tudor Barn*, Eltham. The annual 'Heard/Worked' table was again well-supported with 27 entries, and there were 21 entries for the annual 'G3VLX 40m Challenge': Mike G4AYO won the cup having contacted 107 DXCC entities on 40m during the year. A new Society membership list was issued at the end of the year showing a membership of 149.

1978

In comparison to earlier years in the decade, 1977 had been a quieter year with the Society forced to be more subdued in its activity through legal intricacies. How would 1978 unfold?

There was some contest activity early in the year, but it was all from members' homes – G3RCV was used from Norman G3ZCV's QTH in Orpington for the ARRL CW contest, while Peter G4FUG hosted the ARRL SSB contest activity from his Blackheath QTH.

Moving to new premises
After 15 years of meetings at the Eltham United Reformed Church Hall, the Society moved into new premises at the *Christchurch Centre* in Eltham in January. This move was badly needed as consistently large attendances at meetings had

led to cramped conditions, with many members having to stand at the back of the room, and with insufficient room for latecomers. A well-attended visit to the Crystal Palace TV transmitting station followed, organised by Kevin G8KDC. Members were treated to a tour of the station viewing a remarkable combination of engineering – electrical, electronic, RF, heavy, civil and heating/ventilating, with coax cables like trunking, sophisticated circuitry and self-controlling, automated equipment. Members also saw the 80kW klystron PA which together with the slot fed aerial array gave an ERP of 1MW.

With previous misgivings about Society activities dispelled by its incorporation, G3RCV and G8FCV were aired during an April activity weekend, but member activity was described as *'...disappointingly poor'*. G3RCV/P was active from Wrotham in NFD making 480 contacts, but the entry did not result in a top 10 finish. 15 new members joined the Society in the first three months of the year, including Phil G8OPA and Slim G8OXT (to become G4IPZ).

First limited company Annual General Meeting

The 1st Annual General Meeting of the limited company passed successfully. The meeting was conducted as two separate meetings. The company part of the meeting only transacted formal business and lasted 15 minutes. This was followed by a traditional Society annual general meeting at which recommendations for committee were taken, and at which members were able to raise and discuss matters relevant to the day-to-day running of the Society. Members recommendations for committee were G8LDV, BRS32525, G4FUG, G8KKI, G3VLX, G4DFI and G3SXE. G3JJC remained President, with G2MI and G3ANK as Vice-Presidents. The Board conferred 'honorary' membership to Brad ZL4AD (ex-G3DNC) and Terry ZE2JK. The appointments were ratified by the Board at a subsequent committee meeting. Bernard G8LDV took over as Chairman and I was elected as Vice-Chairman, even though amateur radio was about to take a back seat due to marriage. Bernard was most complementary about *'...the magnificent job'* I had done in the previous three years as Chairman. During my time in the chair the Society had seen some heady, successful times, and a stunning membership peak of 149, which is still a record today.

Another trip to the Purbeck Hills for VHF NFD followed in June. Although the weather was dry, band conditions were poor, with activity reported at a reduced level compared to previous years. However, the Society was active on four bands, including further outings for G3RCV/P and G8FCV/P. The team comprised 19 members: 10th place was achieved.

Special event stations were active in support of the *Greenwich Festival* in June, the *Lambeth Country Show* in July, and the *Victory Social Club* in September. Roy G3JHI again organised the successful GB3LCS activity from Brockwell Park, with three stations running simultaneously, including one into a TH6DX yagi. G8FCV/A was active on 2m from the Hayes event.

The July/August QUA reported that President and QUA editor Stan G3JJC had suffered a 'minor' stroke. Peter G3RZP temporarily took the editorial reins, with Bob G8JNZ typing QUA.

G3RCV/P was active in SSB Field Day from Crayford Marshes, organised by G3VLX and G3RZP. 836 contacts were made using an FT101B, KW1000 amplifier and a TA33Jnr yagi. The team came a creditable 5th. It was noted that the success was as much down to those who fetched, carried, pushed and heaved, as those who operated.The demise of the hitherto successful annual buffet/dance was announced in October. The *Tudor Barn* venue had been criticised as not being particularly suitable, and efforts to find an alternative venue had been unsuccessful. A disco night at the *Christchurch Centre* was suggested, but when tickets went on sale there was little interest.

Also announced in October was the cancellation of an entry into the CQWW SSB contest from the Victory Social Club in Hayes because of a requirement to remove all the antennas by 5pm on the Sunday afternoon! However, a successful time was enjoyed at The Fort in Dulwich when GB8SLS was set up at the Scout Activity Centre to help celebrate JOTA's 21st birthday. Dave G8OHJ (to become G4NOW) organised the 2m event. During 12 hours of activity, an impressive 205 contacts were made, including 141 on FM.

Dave G8OHJ operating at GB8SLS

By the year end, contest activity was noted as having been much sparser. Subscriptions for 1979 had been announced – a £1 increase to £4. New rules for the WCV award were published, and membership had dropped...to 146!

1979

It was reported that nearly all the additional finance from the year's subscription increase had been spent on purchasing a new duplicator to print QUA. Peter G3RZP issued his final QUA in February due to him obtaining gainful employment. The editorship passed into a few hands over the ensuing months. Chris Knight A8996 became the new editor for a few months, with Bob G8JNZ continuing to do the typing. Bob then temporarily edited one issue, before Alan G4BWV (with Bob's typing) assumed the editorial chair.

Seven members took part in the RSGB AFS contest. In those days, teams consisted of five members. The Cray Valley "A" team of G4FAM, G4BXT, G3XRX, G3RZP and G2MI came 3rd. G4FAM was 1st in the overall results. In March, G3RCV entered the BARTG Spring contest, the first time the Society had entered an RTTY contest on an international scale. The station was run from Peter G4FUG's QTH. Eight members took part in the activity and came 19th of 135 worldwide entries.

The first signs of the administrative disciplines of the Society's new status were seen when eight sides of Annual General Meeting paperwork was circulated with the March newsletter. The two-part annual meeting was conducted by Bernard G8LDV, who was re-elected Chairman and I was elected as the Society's third Vice-President. Soon after the meeting Deryck G3VLX resigned from the Board due to other competing commitments on his time; grateful thanks were expressed for the unstintingly work he had done for the Society over many years. Alan G3ANK continued providing Morse tuition between April and October, with three quarters of the class taking and passing the Morse 12wpm test. A new membership list was published in July. Although somewhat reduced from the 149 of two years earlier, it showed membership at a healthy 120. Archive material shows that membership rose again by the end of the year to 132.

The Society took part in NFD from Crayford Marshes site, with G3RCV/P making 468 contacts. VHF NFD was entered from a site at Hastingleigh, near Ashford,

Kent. 23 members took part. QUA reports that the weekend was particularly uncomfortable for hay fever sufferers, and that band conditions were only slightly above average. Even so, some good DX was worked: the 2m station making 401 contacts, the best contact being 731km into Germany. Unfortunately, archive material shows no record of the official results.

A team of stalwarts operated G3RCV/P from Knockholt, Kent in August for the RSGB 70 MHz contest. With poor weather and poor conditions only 53 contacts were made. G8FCV/P was used for the Victory Social Club fete later in the month. SSB Field Day was entered for the third year, but there were issues in gathering a suitable team.

Another activity weekend contest was arranged for October; Margaret G8LXK operating G8FCV on 2m and Alan G4BWV using G3RCV on 10m, but activity was poor. In the not too distant past, contesting had been the mainstay of Society activity and brought Cray Valley to the fore as a successful and active radio society, but those days had faded into the past. Apart from field days, there was now little organised contest activity.

There was some mild QUA editorial mischief with the suggestion that the Society was going through a phase of different factions, particularly when it was announced at a meeting that two members had passed the Morse 12wpm test and had applied for G4+3 licences. Apparently, comments were heard of '...*you can go and sit on the other side of the hall now*'! The Society has always been for the benefit of its members, licensed or not, but looking back now, there had long been mutterings of a G3/G8 divide, with each class of licensee on different sides of the hall, and some Class A members sitting in the same place at every meeting! There was also concern, this time by the Board, that having moved to a larger venue to accommodate increased attendances, numbers appeared to be dropping - even though 'drop' appeared to mean numbers upwards of 50! Formal meeting lectures remained of a high standard, and new members continued to be recruited, including Dave G4BUO, George G3MZR, Cliff G8CKH and Ian Connor (to become G7PHD), while others obtained new callsigns, such as Slim G4IPZ, John G4ILH, Jim G4IJC, Len G4IPF and Alan G4ISC.

The latest Cray Valley SWL contest was arranged to coincide with an international contest weekend and attracted 20 entries from British and European SWLs. A very successful first 'Ladies Social Evening' was arranged for the mid-September meeting at the *Greyhound* in Eltham. 60 members and their YLs and XYLs attended. A raffle was organised, and Bob G8JNZ, his YL Jackie, and Chris G8CIU provided recorded musical entertainment. There was also a darts tournament.

Chapter 5: A decade of ups and downs: the 80s

The 80s were an altogether quieter decade, with perhaps the most noteworthy event being our 40th Anniversary in 1986 when many past and present members attended a specially arranged celebratory evening, but before I look at the celebrations let us look in more depth at how the decade unfolded.

1980

The first event of the year was the RSGB CW AFS contest. Smudge G3GJW operated G3RCV and made 48 contacts. However, there was concern amongst the Board at the general lack of organised activity and that attendances at meetings were falling. An Open Board meeting was arranged for the January mid-monthly meeting to discuss these concerns. Two main 'grouses' emerged:

i)   although being a limited company had its advantages, members did not want constant reminders of company law, and

ii)   there was insufficient feedback on what was decided at Board meetings.

Some members appeared to object to the Board taking decisions which, they believed, should first have been discussed at a Society meeting. An interesting statistic also emerged - 55 members attended the Open meeting, but this was many fewer than at the previous meeting which featured a talk about Metropolitan Police communications. With an attendance of 55, you perhaps wonder now if there really a problem at this time about 'reducing' attendances.

Given the views expressed at the Open meeting, it was perhaps a surprise to read in QUA that the Annual General Meeting was *a very tame affair*. 52 members (under half the Society's membership) attended. Few questions were put concerning the Society's finances, and the various resolutions were passed without question. The informal meeting was held following the formal proceedings, and there was great acclaim when Alan G3ANK was elected President. Fred G3XFG was elected as an Honorary Life member. Stan G3JJC was elected Vice-President. Bernard G8LDV was re-elected Chairman. Alan presented a number of trophies and awards - his first 'official' duty.

Alan accepted the honour of being elected President with *'overwhelming surprise'*. He took to QUA to express his sincere thanks, and said:

"I am very much aware of the responsibility entailed in upholding the high standard of CVRS which has been brought about by combined unstinted efforts of the committee and active members down through the years and all gratefully acknowledged. With your continued support, we should all make sure that the future activities in the CVRS calendar continue to be worthy of the efforts required to maintain the high standard much sought after by other similar organisations and to assist our fellow members in need of help and guidance with their hobby. I know we can achieve our objects. With your help, let us make sure! 73 de Alan, G3ANK".

Bob G8JNZ, who had typed the QUA newsletter stencils during his lunch breaks for three years, announced that he would be unable to continue to do so due to work commitments. Alan G4BWV, editor at the time, was delighted when Eric G3VYM volunteered to take over the QUA editorship.

The April mid-monthly meeting was again held at the *Greyhound* in Eltham, this time because the *Christchurch Centre* was hosting a theatrical presentation, despite the Society's Thursday night block bookings. The social evening attracted 50 members and their YLs/XYLs. Another darts tournament took place with eight three-man teams. In a cracking final, Bernard G8LDV, Shirley (Bernard's XYL) and Dave G8RGD defeated previous winners Cris, Joan and myself in a close finish.

In May, 12 members enjoyed a visit to The Thames Navigation Service of the Port of London Authority at Gravesend, Kent. A number of new members had joined in the early part of this year, including Richard G8ITB. A questionnaire was circulated with the May QUA to obtain members' views on what they wanted from meetings. Only 30% of members returned the questionnaire; one recurring suggestion was to have an HF station operational on natter nites.

Participation in the 10th Cray Valley SWL contest increased, with 28 entries received, including logs from the USA, Canada and Austria.

22 members and 'Zak' the dog took part in VHF NFD in July, again from Hastingleigh. The 70/23 cm mast was 'Cray-Valleyed' due to a faulty joining sleeve. The resulting fall broke the 70cm Multibeam in three places, but it was repaired and ready for use at the beginning of the contest. Conditions were only average, but the claimed scores were up on the previous year. G3RCV/P operated on 23cm making 46 contacts.

Also in July, Roy G3JHI was again the driving force behind another special event station at the Lambeth Country Show from Brockwell Park. It was an event well-supported both by members and the public.

Keith G3TAA and others: VHF NFD 1980

The September issue of QUA made for a particularly dismal read as the editor alluded to a *'…decided lack of contributions this month…'* and Arthur G2MI wrote his last 'HF bands' article because of *'…almost no support from the membership…'*. QUA had been an essential read for many years and the following month I asked, in a hard-hitting article, *'Is this the end of QUA?'*, but QUA that month was a nine page issue! My *'broadside'*, as it was reported by the Chairman, certainly prompted a response…in fact five responses, together with a first 'Chairman's Chatter' column. There was acceptance that an editor's job was not a particularly easy one if there were few contributions, but there was also general agreement that it was time for other contributors to come forward to ensure the newsletter continued.

An entry into SSB Field Day took place from the Society's regular site on Crayford Marshes, with favourable weather for the third successive year. With a rule change to give a bias towards contacting Region 1 portable stations, the team made 800+ contacts.

The Board re-instated an activity contest with some success. 48 logs were received, with a number of Society members active. In November, the main meeting was a visit to British Telecom, London and South East, at Bromley. Martin G3ZAY arranged the visit. He and two members of British Telecom's Public Relations Department gave a lively and well-illustrated talk on Prestel. Prestel, if you remember, offered much in common with CEEFAX and Oracle but it was on its own in the UK by allowing 'talkback' to the machine, e.g. 'talk' to the information providers, order goods from suppliers, ticket information, prices, etc., and pay by keying in a credit card number.

Society President, Alan G3ANK, was reported to have had a heart attack in December, but was heard on the Christmas morning net. December also saw a 12-page QUA; perhaps the message had got through about contributions. The year concluded with an inter-Society quiz with the Gravesend Radio Society and a Christmas meeting in the *Greyhound* (Cray Valley's second home at that time!). The Cray Valley team of Lyell G6HD, Bob G8JNZ and Cris G4FAM were victorious by a close, five point margin. Once again a Cray Valley darts tournament was held, with Slim G4IPZ's team victorious. It seems he also won several raffle prizes (so for those current members reading this, it seems Slim simply has a long-established penchant for raffle success!) However, due to the landlord taking exception to 'an incident', Society members were barred from the establishment!

The Christmas nets were a success, with the HF net run by Peter G4FUG attracting 18 callers and the VHF net run by Tim G8GGP attracting 20 members.

1981

Subscriptions rose to £5 (£2.50 for Associate and Country members), and Bernard G8LDV in his QUA 'Chairman's Chatter' column spoke of the previous January's Open Board meeting at which *'...a lot of foggy mystique that hung around the committee...had been cleared.'* He believed the Society was stronger for the dialogue and that there was a better interchange of ideas and views. Membership stood at 125.

1981 was to be something of a disjointed year. The February QUA was Alan G4BWV's last as editor as *'...it has become a chore and not a pleasure'*, and the *Christchurch Centre* had advised that it would not be possible to hold several main meetings there due to other bookings. The March main meeting and the mid-July meetings were cancelled as alternative venues could not be found, and the April meeting was re-scheduled at the Church Hall of the Holy Redeemer Parish Church, Lamorbey, in Days Lane, Sidcup.

QUA merry-go-round

There is no record of a March QUA, as no immediate successor to G4BWV was found. However, Bill G4KXY took over the editor's job in April *'…with a great deal of trepidation'* having had no previous experience. He was also a newcomer to the Society. As Bill had the time required to cut the stencils, Eric G3VYM's sterling work as stencil cutter was no longer required. Unfortunately, Bill suffered a bad accident setting up the printing machine for the August QUA run, which meant no QUA was issued in August and September. Deryck G3VLX stepped in and was to continue as QUA editor until September 1983, with John G4ILH's XYL temporarily taking on the job of typing the stencils.

Owen G4DFI was elected Chairman at the Annual General Meeting, which was reported to have seen *'…the highest attendance since the forming of the limited company'*. He hoped for continuing success and highlighted the keenness of the new committee to maintain an even meeting programme, as well as trying to rebuild the previously successful contesting side of the Society. New members included Salvador EA8XS (Canary Islands), Erwin VE2FUQ (to become G4LQI), Dave G8ZZK, John G3LNT and Tom, an SWL from Michigan, USA. This ability of the Society to attract international membership shows its global pull at that time.

20 members were involved in VHF NFD in heat wave conditions from Hastingleigh, but there were some *'eventful moments'* – the sudden loss of 70 and 23cms due to the feeder *'dropping off'*, and severe electrical noise on the 2m station on the Saturday evening due to overhead power line noise. Roy G3JHI organised another Society special event station for the *Lambeth Country Show* in July, this time using the callsign GB4LCS. The event went well and without incident.

Million-mile Challenge

A station was set up at the 1st Royal Eltham Scout HQ for JOTA in support of the *"Million-mile challenge"* to raise funds for the RAIBC. GB4RES was aired, with the idea being to contact radio amateurs to achieve a total distance of 1,000,000 miles. To try to achieve such a total, a Western DX33 3-element yagi was erected on a 40' mast for the HF bands and a 6-element quad for 2m. The Scouts maintained their own log and recorded the mileage. A steady stream of Stateside, Caribbean and South American contacts took the total mileage so close to the target, but ultimately, the total fell 70,000 miles short. The event was really well supported by members and the RAIBC benefitted with a £400 cheque, which was presented to an RAIBC representative at the mid-February meeting.

G3RCV/P was active for SSB Field Day in September making 1,295 contacts. An activity weekend was arranged for November during which G3RCV and G8FCV were active. 59 logs were received, an increase on previous years. The Society's 11th SWL contest attracted 23 entries from six countries; additional publicity had failed to secure more entries. The main December meeting saw another quiz – this time against the North Kent Radio Society, where Cray Valley retained its 100% record against visiting Societies. Once again, QUA was not published in December.

1982

Moving into the fifth year of Cray Valley Radio Society Limited, there was delight that the Society had almost turned full circle having enjoyed a number of events and activities in the previous twelve months. Members had come forward to assist, but there was still work to do to encourage other members to get involved with Society activities. The committee (note that QUA was not referring to 'the Board') accepted that having such a large membership with a variety of pursuits, personalities and age gaps required more work to obtain a greater inclusivity.

The 1982 CW AFS contest showed improved participation with entries from G3XRX, G2MI, G3SXE, G4KGM, G4BWS and G3ANK. Unfortunately, it was not a winning entry.

Members of the Reigate Amateur Transmitting Society were invited back to judge the February construction contest. There were three sections i) Best constructed; ii) Most useful; and iii) Novelty, but there were only four entries! Chris G8CIU won the Reigate Cup for his 10GHz SSB transceiver, with Brian G2WI receiving 2nd prize for an HF receiver. It is believed that participation was poor as a result of some adverse comments made by the Reigate judges about certain projects the previous year.

The Annual General Meeting was a protracted affair spread over two evenings mainly because of a marked reluctance of members to put themselves forward for the committee. Eventually, Alan G4BWV, Peter G4FUG, Glyn G8KKI, Andrew G6ALB, Margaret G8LXK, Graeme G6CSY and Brian G4LYU were elected There was no change to the Honorary posts. Arthur G2MI made a number of trophy and award presentations. Chris G8CIU conducted an instant opinion poll at the meeting to try to tease out of members what they wanted of the Society. It produced the following somewhat interesting views:

Meetings should continue at twice a month with at least six natter nites a year. A majority liked a mixture of technical and non-technical lectures, but also appreciated film and slide shows. Nine members offered to give talks.

Construction Contests were favoured by 27 members, but only eight said they would submit an entry. 24 members were in favour (with 14 against) of a Society project.

Contests were generally supported. The majority favoured taking part from home, but 24 were in favour of continuing activity weekends, and 17 supported field days. 23 were in favour of activity in JOTA and 24 were in favour of helping to run a special event station.

Other activities such as quizzes and external visits were popular, but only a small majority wanted to see social events.

QUA was popular. 18 members volunteered to contribute.

Venue was criticised because of poor lighting and acoustics, but 33 members were satisfied, 10 were not.

Other items covered members' preference for the use of 'committee' not 'Board', but an acceptance that some company terms were necessary. All except five members were in favour of retaining limited company status. A large majority thought the level of subscriptions was about right, but eight thought it was too high.

The May meeting was billed a 'Surprise meeting'. With membership above 100, how many could say they knew their fellow members? Derek G3XMD sought to remedy this by asking the 57 members present to say something about themselves. It was reported in QUA as a really enlightening meeting. Also in

May, five members operated from Downe Scout Camp. This included closed-circuit TV and a tele-printer demonstration.

A reminder about 'Society nets' appeared in the May QUA. Indeed, there had been no reference to such activity in the newsletter for many years. It was not, perhaps, surprising that participation had dwindled to just a few stalwarts. With the 2m net in limbo, members were clearly not talking to each other much outside of meetings.

G3RCV had remained inactive since the 1981 Christmas morning 160m net, but it was activated on 23cm in VHF NFD at Hastingleigh. Thunderstorms were the main problem, with the station sustaining some electrical damage. G3RCV/P was also active in SSB Field Day. With fine weather, for a sixth consecutive year, not as many contacts were made as in 1981 but a higher score was claimed. The entry achieved 2nd place. For interest, previous placings had been: 1977 – 29th; 1978 – 5th; 1979 – 14th; 1980 – 12th, and 1981 – 6th.

For a third year, a station was established using the callsign GB4KOS (Kemnal's Own Scouts) in Sidcup. A special demonstration of 70cm high definition colour TV was provided for a short time on the Sunday morning. This proved very interesting and included a transmission back to the station of a videotape made at the site while the demonstration was being set up. Thanks were expressed to G8CIU, G8CTT and G4EGU for this demonstration.

The 12th SWL contest was more successful this year with 39 entries from 11 countries. Owen G4DFI had been administering the contest for a number of years and was grateful for the additional publicity which Tom Land had arranged in stateside newsletters.

The year ended with a membership of 123.

1983

The mood during the early part of the year appears to have been quite downbeat. Even before the Annual General Meeting there was lively debate at an Open Forum about the likelihood of there being no prospective nominations for committee. The point was made that the job was not an arduous one, but that it did require keen and conscientious work. Although members were content with the meeting programme, there was a suggestion that one meeting a month might be sufficient; the suggestion did not meet with general approval. Doubts were also expressed about the suitability of the *Christchurch Centre* as a meeting venue, and members expressed the view that natter nites should not, in general, be used for some other purpose. Arthur G2MI resigned from the Society, it is believed owing to a dispute over the annual 'Heard/Worked' table.

A brighter episode saw 22 entries – possibly a record even today – for the annual Construction Contest. G8AMU, G8AZC and G3BBR from the Reigate Amateur Transmitting Society judged the entries. The Reigate Cup was won by Mason G4MIN for a home-brew keyer. Richard G8CTT was judged second for his home-constructed UHF converters. The 'Most useful' prize went to Graeme G6CSY for a 50W dummy load.

Back to the darker side for the Annual General Meeting, which got off to a less than auspicious start with complaints about the late presentation of the 1982 accounts and the absence of a printed agenda. Phil G4EGU was elected Chairman. The only change in committee was Cris G4FAM's re-election. Members commented that the committee should try to give Society activities more publicity in the amateur press, and were urged to organise visits, special event stations and an activity weekend.

The message from the Chair in the next QUA was that it could be *'…a difficult year'* and that it might not be an exaggeration to describe the year as *'…make or break'*. Fortunately, members rallied and a number of events took place later in the year. However, by the year end, the Society lost 25 members.

Notwithstanding the earlier comments about natter nites, Chris G8CIU came up with the idea that it was better to do something rather than just talk at these mid-monthly meetings. The result was the first Cray Valley DF Hunt. Starting from the *Christchurch Centre*, 14 teams – 30 members – set off to try to find a hidden 2m 'fox'. Transmissions from the 'fox' were every 10 minutes. Although not all were successful in tracking down the hidden station, located five miles away on Chislehurst Common, the event ranked as one of the more successful organised events, with everyone having a good time…and asking for another DF Hunt to be organised. The organisers G8CIU, G8CTT and G3TAA were warmly congratulated for *'arranging something different'*.

A special event station was organised for the *Danson Park Show* in July. A Mosley TA33Jnr yagi was used on the HF bands, and the station housed in a tent loaned by Dave G8ZZK. Members of the public showed a great deal of interest in live colour television pictures transmitted from 'a roving reporter' (Phil G4EGU on camera and Chris G8CIU on the transmitter). The event organiser was also impressed, such that the ATV crew were allowed to film events in the central arena. Although 24 members took part, QUA suggested that for a Society with over 100 members, support could and should have been much greater.

A team entered VHF NFD, again from Hastingleigh. They were not overjoyed to find the site was overgrown with thistles and with clear evidence of recent occupation by a herd of cows! Entry was to the Restricted section, with the 23cm station being located in a separate tent for the first time, using G3TAA's 8' dish and G4EGU's 2C39 linear. Contacts were made into France, Holland and Germany. As this was G3TAA's 21st field day, a bottle of wine was cracked open in honour of the event! The team did well, achieving 3rd place overall. G3TAA/P and G4FAM/P obtained 2nd places on 4m and 70cm respectively.

Peter G3RZP and Lynne G4FNC were married in Swindon in September to become the Society's second licensed couple. News came through that Prince George of Rouillon had passed the RAE with two credits (to become G1CPB, and later G0BDQ). He was to become one of the Society's more eccentric members at this time!

Gale force winds handicapped the SSB Field Day team in setting up for the contest at Crayford Marshes. Two crank-up towers were eventually erected with a 20m quad and a 3-element DX33 yagi, plus two loops for 40m and inverted-V dipoles for 40 and 80m. There was, however, a reduced operating team as at least four core contesters were unavailable, but G3RCV/P made over 1,000 contacts.

Cris G4FAM at SSB Field Day in 1983

66

Norman G3ZCV ran Morse classes through the autumn at the Adult Education centre at Eltham Hill School. Together with Dick G8HBM, Norman also taught the RAE at the same establishment. Slim G4IPZ took over the editorship of QUA in October from Deryck G3VLX, and GB4RES was aired from the District Scout HQ in New Eltham for the October JOTA weekend. The Cray Valley SWL contest again proved its popularity amongst short wave listeners with 38 entries from 14 countries, including first time entries from Australia and Russia.

A Christmas dinner was held at the *Eden Park Hotel*, but on arrival nearly 40 members found it was actually a *Mr Toby's Carving Room*! However, the meal was reported in QUA as *'splendid'* and the price also included a free liqueur. However, to much surprise, a cup of coffee was an optional extra. Some members were mildly irritated by this additional expense, so it appears that Cris G4FAM raided his wallet to buy coffee for those who did not want a liqueur coffee!

Considering the seemingly problematic first few months of the year, the Society had enjoyed an active and successful year.

1984

The sad and sudden passing of Peter G3WVP from a heart attack was announced in March. He was 52 and was a well-known figure within the Society. He was always willing and eager to help, especially the blind, the handicapped, and the less privileged. He operated JOTA and other exhibition stations to promote the hobby. He was responsible for the successful implementation of the 'Worked Cray Valley' award – an award which gained international recognition. He also held office within the Society and was a staunch supporter of the Radio Amateur Invalid and Blind Club. On a more pleasant note Brian Rowe and Frank Parradine passed the RAE. They were eventually to become G4WYG and G0FDP.

The Annual General Meeting was moved from the main meeting in April to the mid-monthly April slot (where it has been ever since) to allow sufficient time for the necessary paperwork to be made available to members. At the meeting itself, there were, controversially, no nominations for committee. Members recommended that the previous year's committee should be re-elected, but with Graeme G6CSY becoming Vice Chairman. These recommendations were ratified at a subsequent committee meeting. However this and other concerns, led to an Extraordinary General Meeting being called later in the year. On a more pleasant note, Alan G3ANK, who had moved to Christchurch in Dorset, presented 'The Founders Cup' to the Society, to be awarded to recognise *'long and meritorious service to the Society'*. The first recipient was Derek G3XMD.

The Society callsign was used for a Scouting event at Downe in May. An FT101ZD transceiver and a linear amplifier provided about 350 watts to a 5-band vertical. The highlights were Derek G3XMD operating in full dress Scout uniform, and another appearance of the Cray Valley portable amateur television team.

Building on the success of the first DF Hunt the previous year, the 1984 event was also a success. Amazingly, it attracted 27 competitors in 12 teams. Starting from the Eltham meeting place, teams drove around the area taking bearings to locate the 'fox' which was hidden in Sidcup, Kent.

First real signs of apathy and concern

"Rifts, apathy and complacency leading to the folding of the Society" were strong words used in QUA later in the year to voice a concern held by some members following an Extraordinary General Meeting – of which there was no report in QUA – called to ratify the actions of the committee since the previous year's Annual General Meeting, and express concern that there were no nominations for the 1984 committee, and that membership was falling. The editor expressed the view that apathy and complacency did exist within the Society, but he believed there were fewer rifts than 5-6 years earlier, when there was a distinct *'them and us'* situation with regard to class A and class B licence holders. Indeed, I have already referred to the imaginary line drawn at meetings, when class B licensees sat on one side of the room and class A licensees sat on the other. Falling membership was taken as an indication that the Society was not fulfilling a need. The crux of the issue was that more members needed to volunteer and get involved with Society activities to address the pessimistic tones that were emanating from the Society.

The stalwarts put on a further station at the *Danson Show*, but there was a depressing lack of participation from Society members. There was also a lack of interest in taking part in VHF NFD, but Keith G3TAA and Graeme G6CSY operated from the Hastingleigh site on one band to ensure that Cray Valley had a presence in the results.

SSB Field Day success

G3RCV/P first entered SSB Field Day in 1977 with a score so low that it was best forgotten, but the team improved steadily to gain second place in both 1982 and 1983. Further work to improve antennas and greater concentration on working multipliers rather than just 'running' led to a first SSB Field Day success and the top spot in the Open Section of SSB Field Day in 1984. Unfortunately, there is no detailed report of the contest in QUA, but the editor congratulated the successful team and those who helped on site before and after the contest.

A September activity weekend saw nine entries, with Vaughan G4MVR achieving the best Cray Valley score. It was also announced in September that 'The Peter Vella Memorial Cup', donated by Peter's widow Kath, was to be awarded annually to the member who contacted most Cray Valley members, Society stations and special event stations organised by the Society during each year. This award reflected Peter's tireless work to introduce and make a success of the 'Worked Cray Valley' award, which encouraged members to talk to each other. Also in September, Slim G4IPZ announced his intention to stand down as QUA editor. Graeme G6CSY stepped in temporarily until the end of the year to ensure QUA was published.

Members attending meetings in the 80s paid a 20p 'entrance fee', which to commonly held belief paid for 'half time' tea and biscuits. Wrong! The money was used to pay for the hall rent, which had risen twice in quick succession. From the main November meeting the fee (known as 'bag money') rose to 30p. Interestingly, the point was made that this increase applied to ALL members!

QUA reported that 40 members, relatives and friends returned to the *Eden Park Hotel (Mr Toby's Carving Rooms)* for Christmas dinner.

1985

The QUA 'carousel' turned full circle with Deryck G3VLX taking back the editor's job for the January issue, but he warned it would only be a temporary measure. And so it proved, with Vaughan G4MVR appearing as editor in April.

Two of the year's early meetings had to be cancelled when the hall owners failed to inform the Society that the hall was booked for other purposes; a very unsatisfactory situation. This, and ever-worsening membership numbers, led the committee to discuss finding an alternative meeting venue.

Frank G3WMR judged the 1985 Construction Contest. He awarded the Reigate Cup to Chris G8CIU, whose winning entry was a 10GHz receiver, transmitter and dish assembly. The overall entry was down on previous years, but the standard very good. GB0LBK appeared in April to mark the closing of Littlebrook Power Station. The event was organised by Steve G6SDO.

A small group of members attended the award presentations at the RSGB National Exhibition in Birmingham to collect the *Northumbria Trophy* for winning the 1983 SSB Field Day. Instead of the usual cup, the trophy consisted of a sphere topped by a loaded whip, mounted on a handsome marble and gilt base. The team could not take the trophy away, but were photographed with it. A small plaque, for the team to keep, was also presented, but the inscription was mounted upside down!

There is no mention in QUA of proceedings at the 8th limited company Annual General Meeting, except that Alan G3ANK presented the trophies. Kath, widow of G3WVP, presented the 'Peter Vella Memorial Cup' to Charles G4DNR, who had been severely disabled since a car accident (in which he was the innocent victim) 25 years earlier. As the cup was presented, it was with tears in his eyes that Charles said it was the first time he had been able to visit the Society since joining in 1972, and what a happy occasion it was for him. It was with great sadness, therefore, that members heard of his passing the following Wednesday.

After the meeting Arthur G4BWS left the hall for the last time before his move to Burton-on-Trent. Why does that get a mention in the Society's history? Because Arthur had been tea-maker-in-chief to the Society for many years since he joined in 1973. It is behind-the-scenes jobs, such as making the tea, which are often overlooked but which are so important to the smooth running of meetings at any amateur radio society.

Brian G4WYG began CW classes on natter nite evenings in June. It was also announced that an 80m net was to be introduced to replace the long-running 160m net. This change was made with the Society's many country members in mind so they could keep in touch with Society. Eagle-eyed members noticed a heading change on page 1 of the July QUA; even sharper-eyed members noted a change in callsign – from G8FCV to G1RCV.

The *Danson Show* in July, using GB0DAN, was reported as the most successful special event station to have been organised by the Society to that point. Stations were established on 2m and on HF, with mobile 70cm ATV proving as popular as in previous years. QUA reported '...*uncommonly large numbers* of members (29) attending the event.

Cray Valley members at GB0DAN
L-R: Unknown, G3VLT, G3TAA, G4EGU, G8DYN, G8CIU, G8CTT, G4DFI, Bob Francis BRS88021, G4IPZ and G6PKS

New meeting venue secured

As referred to earlier, for one reason or another, the time came to look for a cosier meeting venue. One was secured, but only if the members liked it. Graeme G6CSY and myself viewed the hall and agreed to give members a say in the decision-making. The mid-August meeting was arranged for Progress Hall in Eltham, one of the St. Mary's Community Complex group of properties. 40 members attended the meeting. Following a show of hands towards the end of the evening, members agreed that the application to use the venue should be pursued. The first meeting at the hall was confirmed for mid-October. The Society would use these premises for meetings for the next 26 years.

Despite doubts about the weather, the 2m 'fox' hunt went ahead as planned. The 'fox' was hidden in Blackfen, Kent. Seven teams took part, five of which found the 'fox', before adjourning to a local public house to discuss the finer points of Direction Finding. The Society did not take part in VHF NFD, but G3TAA and G3VLT used the Hastingleigh site on 4m only making 179 contacts.

More uneasy times

The August QUA bought into the open certain ex-members' dissatisfaction with the Board and details of why they had left the Society. Much centred on the Board's apparent failure to handle aspects of the 1983 and 1985 Annual General Meetings in accordance with the Articles of Association, and the failure to report the results of the 1984 Extraordinary General Meeting to members. With a reluctance of members to take office when the words 'limited company' applied, it was asked whether the time had come to stop being a company limited by guarantee. However, what had been forgotten was that a huge majority of members democratically voted to accept the proposal to become a limited company following lengthy investigations and a public treatise from a respected member on the benefits of assuming such status. Calls were made for the membership to support the Board and pull together, with a view expressed that any proposal to remove the limited liability by winding up the company would be strongly resisted. That proved to be the case for a further 12 years.

Following the SSB Field Day success in 1984, G3RCV/P was active again for the 1985 contest, but band conditions were reported to have been unfavourable, although the number of 'multipliers' (British and European portable stations) were higher than the previous year, as was the claimed score.

G3JJC SK

The December QUA reported with great sadness the passing of Stan G3JJC, a real gentleman. Stan joined the Society in 1953 and had been President from 1969 to 1978. He had also been Vice President in 1968/69, Chairman from 1964 to 1966, Secretary from 1963 to 1964, and a committee member in 1960/61 and from 1966 to 1969. He was also QUA editor for many years, providing a monthly read of consistently high quality. I recall visiting him on numerous occasions during the 70s and 80s to provide 'copy' relating to SWL and DX news. Stan was certainly remembered fondly by his many amateur radio friends. The Society was a beneficiary in his will.

And so ended a turbulent year, where fortunes had ebbed and flowed as never before! With the Society's 40th anniversary on the horizon, the committee hoped 1986 would steer a straighter course.

1986

Owen G4DFI took over the Chairmanship of the Society at the end of January as pressure of work and domestic reasons forced Graeme G6CSY to stand down. Our friends from the North Kent Radio Society were welcomed for an inter-society quiz, but the Society lost its 100% home record, being outplayed on the evening by a total of 33 points. The March Construction Contest featured eight entries. Richard G8CTT was awarded the Reigate Cup for his 24cm TV receiver.

In readiness for the DF Hunt (and future DF Hunts), Nigel G1BUO donated 'The Tally Ho! Cup' to be awarded to the person or team who successfully found the 'fox' first. The first winners of the cup were G3KRW, G0FDP and G0ARW, who found the 'fox' in Danson Park within 43 minutes of setting out from Eltham. Society members visited the Science Museum and saw the GB2SM station in May.

Chris G8CIU was elected Chairman at the Annual General Meeting, with Owen G4DFI as Vice-Chairman. Other elected officers were Brian G4FUG and Glyn G8KKI. Committee members elected were – Nigel G1BUO, Cris G4FAM and Keith G3TAA.

GB2GF was the special callsign used at the *'Plumstead Make Merry'* as part of the Greenwich Festival celebrations. Operation was mainly on 2m. The *'Make Merry'* committee printed 200 QSL cards free of charge, but what was not known until the day was that they had simply copied the 'paper and paste' specimen rather than arranging for printed QSL cards! Even today, the card is unusual, and quite a rare 'amateur' QSL card.

The GB0DAN station operating from the *Bexley Show* was again a success, with a good Society attendance and plenty of interest from the public. Activity was on the HF bands and 2m. There were also ATV stations on 70 and 24cm: the 24cm station being a fixed link from the arena to the GB0DAN operating tent using a very low power transmitter and a large parabolic dish. A professional colour TV camera was used, with the intention of recording the arena events for Bexley

Council as well as transmitting them back to the GB0DAN operating tent. The Mayor of Bexley was very impressed when a contact was made with DK0TA, operating from Bexley's twin town, Arnsburg: local German newspaper *Westfalenpost* carried an article and photograph of the contact.

40th anniversary month

October saw a wonderful month of activity and celebration. A special 12-page 40th anniversary issue of QUA was published, and GB4OCV was active throughout the month: Brian G4WYG on HF and Owen G4DFI on 2m. Several activity periods and impromptu Society nets took place to help members and other radio amateurs amass sufficient points to claim a special 40th Ruby anniversary award: a specially engraved plaque. Indeed, it was with some satisfaction that 55 award claims were received.

The first October meeting saw over 50 members attend a surplus sale which netted the Society a £38 profit. The mid-monthly meeting saw an anniversary party which attracted 64 members, ex-members and guests. Alan G3ANK reminisced about the early days of the Society. This was followed by a buffet and cake-cutting ceremony by two founder members, Geoff G2CXO and Alan G3ANK.

Geoff G2CXO and Alan G3ANK with me at the 40th anniversary party

Geoff G2CXO wrote:
"I must tell you how delighted Mary and I were the other morning to find the October edition of QUA in our mail and to see again that very familiar logo of 'QRM'. Mary was especially pleased since in those very early days of the Society

she and I used to spend hours on our knees on the lounge carpet, spraying coloured ink with toothbrushes through a stencil of the logo in order to produce the front covers of the early editions of the magazine. The stencils had been laboriously and painstakingly cut out of sheet metal by G3MZ, a labour of love if ever there was one! Little did we think at that time that the cover would again be in circulation 40 years later!"

Operating GB4OCV on HF was the first time Brian G4WYG had run a special event station. With some trepidation, he called 'CQ' on 80m. His first contact was with G8RZ in Workington, but the calls came thick and fast, with Brian making a total of 363 contacts.

Chris G8CIU, who had recently obtained his G0FDZ callsign, wrapped up the anniversary month in QUA with his own personal highlights. He had obtained enormous pleasure listening to Alan G3ANK's talk on the early years of the Society, and noted the great success of the anniversary party expressing his thanks to Nigel G1BUO and I for arranging the evening. He said he would remember for many years the *'amazing'* nets on 80m and 2m handled so ably by Brian G4WYG and Owen G4DFI respectively. The success of the Ruby Award stimulated activity amongst members with many 'rare' callsigns contacted. Lastly, Chris did not forget the members, because without their support the month would not have been the success it undoubtedly was.

How did members follow that? Well, also in October members organised two stations for JOTA. GB4RES appeared from the 1st Royal Eltham District Scout HQ in New Eltham, organised by Derek G3XMD, and GB8RE was active for the 8th Royal Eltham Scouts at a separate venue in Eltham run by Dave G4NOW. Chris G0FDZ set up a 70cm ATV link and a successful contact was made between the two stations. During the contact one Scout Leader appeared at GB8RE and then drove to GB4RES and also appeared from there, to be seen at both ends of the same contact.

1987

The year started with only 68 paid-up members, a far cry from the 149 peak of 10 years earlier.

January saw the return leg of the inter-society quiz with the North Kent Radio Society. A victory by 80 points to 49 reversed the result from the previous year. The Cray Valley team was Keith G3TAA, Cris G4FAM and Vaughan G4MVR. Early year joiners were Peter G0GIR, Andrew G1VOH, Fred G0CSF and Dave G8RGD, but the Society lost Richard G8CTT (who had recently been licensed as G0FEB) in tragic circumstances.

The Annual General Meeting elected Brian G4WYG as Chairman, with Chris G0FDZ as Vice-Chairman. Alan G3ANK presented a number of trophies and five 'Worked Cray Valley' awards, including both Gold and Diamond awards to yours truly.

For the fifth year running, Cray Valley set up for the *Bexley Show* at Danson Park. The HF yagi had not performed too well in earlier years, so 10 members went to the park a few days before the event, together with the assembly instructions, to piece it together. The instructions said that all dimensions were critical to something of the order of a millimetre, so with a steel measuring tape, it was laid out and assembled. Once it was assembled correctly, the group stayed to listen to some military style music and fireworks, but they were not prepared for the army of gnats and midges that descended around sunset! The pre-planned antenna assembly paid off, as the HF station made over 150 contacts during the weekend.

Back in 1983, Chris (then G8CIU) arranged the first Cray Valley DF Hunt with Keith G3TAA. Since then, hiding places had been chosen that were accessible by road. But for 1987, something different...the DF Hunt was on foot! A suitable hiding place on Chislehurst Common had been found – a holly bush with plenty of dark, prickly leaves! It was a damp, drizzly evening, ideal for midges and gnats. 19 members took part but only four teams (nine members) found the 'fox' – Keith G3TAA – situated behind the holly bush and under an umbrella!

The Society manned a station at the *'Plumstead Make Merry'* for a second year. GB2GF was active on HF and VHF. Band conditions were generally poor, so contacts were not easy to find. Several unusual occurrences were mentioned in QUA:

i) operators were asked on six different occasions if they were the organisers;

ii)   on two occasions if a call could be put out for missing children; and, most amusingly,

iii)   being asked by a five-year-old if the operator could contact her Daddy, who was a mini-cab driver!

Also in June, a group of eight stalwarts set up a small station on 2m and 24cm ATV for the Hurst Road School fete. The under-used G1RCV callsign was aired. The small team enjoyed the day, as did the pupils. Although nothing was being sold, the day raised over £2,000 to provide extra facilities at the school. GB4AAD was a third special event station to be aired in a short space of time – this time at Biggin Hill Airport for the 1987 Air Activities Day.

Yet another special event callsign was GB0GAS ("Go Air Scouting"). This was aired for the 57th Greenwich (Air) Scouts whose scoutmaster was Cray Valley member Ian Ford. The Scouts and cubs turned JOTA into a camping weekend at the Greenwich District site at Luxted, Downe. The weekend followed 'The Great Storm'; the team drove to Downe past the devastation of fallen trees, but an HF station was on the air before lunch to contact 23 official scout stations in 14 countries. A 2m station was on the air for a few hours on the Sunday morning.

To give it a place in Cray Valley history, I mention 'The QSL Bucket'. This was a 'service' which Ted G3DCC provided. The bucket was at the meetings which Ted attended. Members took outgoing QSL cards to the meetings, placed them in the bucket, and Ted packaged them up and sent them to the RSGB bureau, saving members the postage. As member activity was quite high through the late 70s and early 80s, it was not a surprise to see the bucket overflowing with QSL cards.

The Christmas Dinner and the mid-December social evening were both held at *The Jolly Fenman* at Blackfen. Membership had risen to 74 by the end of the year, but over the next 10 years membership was to fall to its lowest for more than 30 years. Also at the year end, Vaughan announced that his tenure as QUA editor was to come to an end the following April. Who would take over the editorial pen now?

1988

Board changes were announced in January when Peter G4FUG resigned as Secretary, but he remained on the Board until the Annual General Meeting. Brian

G4WYG took over as Secretary. Chris G0FDZ became Chairman, with Nigel G1BUO as Vice Chairman.

A further inter-society quiz with the North Kent Radio Society was a very close affair, with the home team winning by the slender margin of just two points. The home team comprised Sam G4OHX, Bob G8JNZ and Cris G4FAM. Quote of the evening came from Sam when asked the following question: *"In CW, what do you send if you make an error?"* He replied *"I don't"*!

A highly successful event took place in April at the National Scout Camp at Downe. G3RCV was aired on the HF bands and the Scouts were delighted when radio amateurs in the USA were contacted. Chris G0FDZ and Phil G4EGU also wandered around with their portable 70cm ATV station.

Bob G8JNZ became the latest QUA editor in May – a job he was to hold for 13 years. Paying tribute to Vaughan's time in the editorial chair and *'....steadfastly keeping this worthy tome on track'*, he underlined the time, dedication and professionalism required to produce works of literary art every four weeks. Bob spoke of the need for members to offer articles for publication so he could provide *'splendid value to members by providing regular, quality, eight page QUAs'*.

The Annual General Meeting is reported to have been extremely well attended. Chris G0FDZ and Nigel G1BUO were re-confirmed as Chairman and Vice-Chairman for the year. The Board of Directors was confirmed as G0FDZ, G1BUO, G8KKI, G4WYG, G4NOW, G6WRP, G3SXE and myself. Alan G3ANK presented the various trophies and awards, and Richard G0FEB's parents presented the 'The Richard Buckley Memorial Trophy' to Barbara Rowe (G4WYG's XYL). The trophy would be awarded annually at the discretion of the Board/committee for meritorious service to the Society.

Is this the end of Cray Valley special event stations?
A paper was circulated to members with the May QUA which set out *'the rules'* for future contest and special event station activity. The paper concluded *that '...if these rules are followed we can expect to be involved with many successful and trouble-free contest and special event stations'*. However, soon after this paper was circulated, the committee reviewed participation at special event stations due to a concern that support for such events had fallen. A questionnaire

asked if there was sufficient resource and manpower to be able to set-up and operate such events, or whether the time had come to 'call it a day'?

The station at the *Bexley Show* did go ahead on HF and VHF even though the price charged by Bexley Council had risen steeply. The Society's new Cushcraft AV5 HF vertical antenna was used into a Heathkit SB104 transceiver. On VHF, an Icom IC211 and R7000 scanner were used into a Jaybeam 8-element yagi at 20'. Thanks to a kind donation by Sam G4OHX, the Society inherited its first tent for special event stations and field day style events.

Following members' comments about the future of special event stations, and deliberation of them by the committee, it was announced that the Society would not in future participate in the *Bexley Show*, or any other event which charged a fee for taking part. Instead, it was felt that the Society should concentrate its efforts on attendance at free events and in trying to attract more local press coverage of its activities. This really was the death-knell for Cray Valley special event stations, with few being arranged in the next 11 years.

Although not an event supported by the Society, Deryck G3VLX organised GB2MVF from the Marden Village Fete, and Pete G0HUM and Dave G0IMU operated GB75BPW in support of Bexleyheath Police Week. It was thought that the station activated by Pete and Dave may have been the first ever amateur radio station to operate from an active police station within the Metropolitan Police District. It was run entirely by serving or recently retired police officers and the opportunity was used to raise awareness of crime prevention in the shack/home.

The German amateur radio society located in Arnsberg (Bexley's twin town) was awarded honorary membership of the Society in June. This meant that a contact with club station DK0TA would count towards the 'Worked Cray Valley' award. Following the success of the walking DF Hunt the previous year, another was arranged at a squelchy and gnat-infested Chislehurst Common.

The Cray Valley quiz team, this time consisting of Ted G3DCC, Pete G0GIR and Bob G8JNZ, lost gracefully in mid-August to a team from the Civil Service Amateur Radio Society. The meeting had to be moved to a hall at Lionel Road, Eltham due to circumstances beyond the Society's control.

British Vintage Wireless Museum visit

Nigel G1BUO arranged a successful evening visit to the British Vintage Wireless and Television Museum in East Dulwich in October. The party were overawed at the sheer number of radios in Gerry Wells's collection.

On walking through the door, members gained an instant impression that this was no ordinary middle of terrace house, with a line of floor-standing domestic radios at the entrance. The museum consisted of two buildings of 10 rooms with 1,300 wireless receivers on show, together with display cabinets of components and

Cray Valley members at the museum

wireless associated artefacts. Gerry's collection contained a wide range of radios, televisions, speakers and radiograms from the dawn of radio to the last valve models ever made. Not only did members come away having seen some real items of radio of history, but Gerry obliged with practical advice about members' own antique radios, how to repair them, and where to get spare parts.

One of the main aims of the committee during 1987 and 1988 had been to put the Society on a sound financial footing. Resulting from decisions taken at the Annual General Meeting to institute budgetary control on incomings and outgoings, a review found that trading losses over the two years could not continue and that the Society needed to address the steady erosion of its reserves. The committee therefore budgeted for all foreseeable expenditure in 1989 and quickly saw that membership rates had to be increased. Discounted rates for early payers were reduced and discounted country membership was withdrawn. A full year's subscription was confirmed at £7.50 (£7.00 for early payment); Country membership £5; Family/Associate/Senior Citizen membership £3.50.

1989

The year began with the sad news of the passing of Norman G3WJK. Norman joined the Society as a short wave listener in 1963 and had been a loyal member, often to be heard on Cray Valley 160m nets. Prince George of Rouillon G0BDQ also passed away, knocked down by a vehicle while crossing a busy road near his home.

The Society was invited to operate a special event station for the Eltham and Mottingham Girl Guides and Brownies for *'Thinking Day on the Air'* in February. The callsign GB3GEM was obtained. The event was extremely well-supported by the Guides and Brownies, with 33 passing a 'Greetings Message'.

The Annual Construction contest was held in March, with Pete G0GIR awarded the Reigate Cup. The Society was soundly beaten by the North Kent Radio Society in the latest in a regular series of inter-society quizzes between the two societies. Owen G4DFI was awarded a Diamond 'Worked Cray Valley' award, for obtaining 100 points contacting Society members, Society and special event stations. All his contacts were made on 2m.

Chris G0FDZ and Nigel G1BUO were re-elected as Chairman and Vice-Chairman at the Annual General Meeting, but Brian G4WYG stepped down as Secretary due to work commitments which would take him to Sudan. Ted G3DCC became caretaker Secretary.

GB2GF was activated again from Plumstead Common for the *Greenwich Festival*, with 90 contacts on HF and VHF being made, and another walking DF Hunt was arranged on Chislehurst Common. This time, the weather was more favourable, but Chris G0FDZ and Keith G3TAA, who were the 'fox' had to contend with gnats and midges yet again! Frank G3WMR and Dave G4YIB were the first to find the 'fox' and received 'The Tally Ho! Cup' at the post-DF Hunt social. QUA reported the *'merry swigging of amber nectar'*, which was enlivened when Nigel G1BUO, just back from a holiday in Goa, produced a pot of home-made chutney concocted from a recipe he had brought back from his holiday. 'Lethal' was the widely held view!

The mid-August meeting featured 'G3RCV on the air'. The evening was better supported than had been expected. A good evening ensued once a few teething issues had been resolved, including adding an extra length to a hidden long wire, which had been inadvertently 'pruned' by a local resident!

The Society organised its final special event stations of the year when it activated GB4LBD and GB3RES in October. The Society was invited by the District Commissioner to run a JOTA station for the Royal Eltham Scouts. This was reported as *'a flop'* on two counts; the scouts were conspicuous by their absence, and the event was poorly supported by Society members. Only two cubs and two Guides and four Society members attended. GB4LBD was on the air from Littlebrook "D" Power Station in Kent to celebrate 50 years of generating electricity from the site. Antennas were erected by the hosts. The station was thoroughly aired and there was not an electricity meter in sight! Operation was on HF, 2m and some microwave TV. Steve G6SDO did much of the arranging, but Society attendance was poor, relying on the usual stalwarts. However, the large numbers of visitors made the event a success. Steve was to follow up this event with a knowledgeable talk on Power Generation and an organised trip around the power station in 1990.

Once again, the Civil Service Amateur Radio Society inflicted defeat at the inter-society quiz, winning on the very last question. Adjourning to the *Jolly Fenman* at Blackfen, it seems a *'...totally unjustified outburst from the barman'* resulted in the Society having to find an alternative venue for its 1989 Christmas festivities!

Chapter 6: A decade in the doldrums: the 90s

The 90s can be summed up quite succinctly: good programme, generally poor attendances (averaging between 12 and 15), general apathy, no contest activity, and only a couple of special event stations. It was also the decade that saw the Society lose its limited company status, and begin planning for the biggest special event station since GB2LO in 1968. Although there is sparse activity, the decade is worth a closer look.

1990

The first QUA set the tone for the year. The editor wrote *'...Please let me have something for QUA over the coming months...otherwise QUA will just be one page with the CVRS calendar on it'.*

The Annual Construction Contest was quite well supported, with a decent collection of projects to be judged. Dave G0JBT was awarded the Reigate Cup for an autodial tester. Lee G7EHA won the Novice award for home-constructing a Howes 80m receiver. The annual quiz with the North Kent Radio Society was *'...a rip-snorting event'.* Nigel G1BUO was the quizmaster and the questions were from the latest version of *Trivial Pursuit*. The final score was quite emphatic – Cray Valley 147: North Kent 86.

Steve G6SDO followed up the 1989 GB4LBD special event station with a fine talk about Power Generation. He arranged a visit to the power station in May, which was recorded as a *'monstrous success'.* There was so much to see: plant, machinery, videos, and a buffet. It was reported in QUA that most of the party were still there approaching midnight!

The Annual General Meeting passed off relatively quickly, thanks to some detailed, advanced planning. Nigel G1BUO was elected Chairman, a post he was to hold for 10 years. Alan G3ANK awarded the trophies and awards, but there were comments about the complexities of choosing the winners with fewer members interested in competing for them. The apathy theme ran consistently through QUA for the year and in his column following his election as Chairman, Nigel said *'...Well, after all my jumping up and down I ended up as Chairman after all. I vowed I wouldn't do it, but in the end there were no takers and, rather than see the Society fold, I stepped in.'*

Several events did take place, including the Annual DF Hunt; the 'fox' being located at Footscray Marshes. The first three teams to locate the 'fox' were not Cray Valley members, so the fourth team home – G4DFI, G8KKI, G7EHA and SWL Bob Francis were awarded 'The Tally Ho! Cup'. The annual one day event at Plumstead Common as GB2GF took place but it was not a success. Ted G3DCC and Owen G4DFI manfully operated the 2m station for eight hours. Few members supported the event and those who did were not interested in operating. No members were contacted either. So, GB2GF was consigned to the history books.

The September QUA reflected the apathy. The editor said *'...not much to talk about this month. The August Society meetings were extremely quiet affairs with only half a dozen people turning up to put the station on the air or have a ragchew at the Natter Nite and there's been no contributions to the Newsletter, other than from the regulars – plus I can't think up any controversial viewpoint to stimulate your thinking cells...'*

October saw GB8RE activated for the second time for the 33rd annual JOTA weekend. Once again set up at the United Reformed Church Hall in Eltham, the station was manned by Dave G4NOW, Steve G6SDO and Neil G6PXI. Operation was on HF and VHF, and attracted 150 visitors including a complete Air Cadet squadron from Hayes School. Chasing quality rather than quantity, the best contact was with the Greytown Scouts in South Africa, ZS5GSG. As a result of the weekend, Dave G4NOW ran a Communicator badge course for six of the keener Scouts, and all the scouting participants received the official World Scouting JOTA patch to wear on their uniforms. This was reported as a major breakthrough in achieving wider recognition of the JOTA event. The GB8RE station details were sent to Gilwell Park for it to be included in the UK official JOTA report, and a request was made for a repeat event in 1991 to coincide with the group's "Outreach 2000" project, aimed at putting scouting into the second Millennium.

The Society's annual SWL contest received 15 listener entries. This was an improvement over previous years; the shorter 'sprint' type event helping to swell the number of entries. Nobby G7GWG (now G0VJG) joined the Society in November. With differences resolved at the *Jolly Fenman*, a Christmas social evening was arranged, with a free buffet included.

At the year-end membership had fallen to 54. As an incentive to introduce new members, a voucher scheme was introduced for 1991: anyone introducing a new member received a £1 discount, either off of the 1991 subscription or against items purchased at a surplus sale.

1991

'Half 'n Half' natter nites (one half video, slide show, etc. and one half 'ragchew') were introduced in an effort to make mid-monthly meetings more appealing to members. The committee decided to scrap inter-society quizzes due to poor support from members.

Peter G0HUM staged another special event station, this time in the Charge Room and cells at Sidcup Police Station for Crime Prevention Week using the callsign GB0CPW. Although not a Cray Valley event, a number of the operators were Cray Valley members.

The number of entries for the March Construction Contest was extremely disappointing, with only one entry received. However, a good attendance was noted for the 'Half 'n Half' mid-monthly at which two RSGB videos were shown: about the new Novice Licence and an introduction to amateur radio.

A visit to the Amberley Chalk Pits Museum was arranged for a Saturday in May. It was reported as an interesting day out, but the disappointment was that many of the people who had expressed an interest in the visit did not turn up! The museum was set in 36 acres of chalk quarry and was rather like stepping into a time warp with a village blacksmith's shop, a boat-making workshop, lime kiln, pottery and many other views of days gone by. Of course, the main reason for the visit was to see the wireless museum. There were exhibits of radio equipment dating back to WW1, domestic wirelesses, examples of crystal radios, military paraphernalia, and a reconstruction of a Lancaster Bomber radio operator's position.

Nigel G1BUO and I were the DF Hunt 'fox' in 1991. This saw a return to a mobile event rather than the successful walking hunts of earlier years. Using G1RCV/P, the 'fox' was found by eight of the nine participant teams parked in a side road in Bexley Village. The only mistake made on the evening was arranging the post-DF Hunt drinks for an establishment which did not allow females!

Members were saddened to hear that Brian G2WI had passed away in August. He was a member between 1965 and 1983. His voice was well-known on 160m for his 'Wigwam Net' which outlasted all other local nets. Brian left the Society in 1983 to help form the Darenth Valley Radio Society.

The October 'Half 'n Half' meeting was a small, low key, soiree in celebration of the Society's 45th anniversary. A slide show was organised by Bob G8JNZ and me, followed by the usual ragchew, but this time over a glass of wine. Past members of the Society were invited, and GX3RCV was aired for part of the evening.

The following weekend Dave G4NOW organised another JOTA outing for GB8RE on behalf of the 8th Royal Eltham Scouts from the United Reformed Church in Court Road, Eltham. In November, a joint visit with the North Kent Radio Society was arranged to the Shepherd Neame brewery in Faversham. It does not need me to say it was a most enjoyable visit!

Congratulations were due to Lee G0NQB for being runner-up for the RSGB's Young Amateur of the Year award. The year closed with a poor turnout at the Christmas Social at the *Jolly Fenman* and on the Society's Christmas nets, but in terms of membership, stability had been achieved.

1992

The poor meeting attendance theme continued into January, when only seven members attended an RSGB tape/slideshow of a Mexican DXpedition. As a result, the committee decided to cancel the first main meeting in January until further notice. Both February meetings had to be re-scheduled when illness and family commitments prevented the speakers from attending. As an alternative for the second cancellation, Dave G0JBT showed the wonderful Tony Hancock video, 'The Radio Ham'.

Cray Valley Junk Sales (re-badged a couple of years earlier as upmarket 'Surplus Sales') were renowned for bargains, and the proceeds of the mid-January sale were donated to the RAIBC. This sale was particularly notable for the appearance of the 1948 edition of a book on how to convert various war surplus sets for amateur operation. The audience were somewhat stunned when it was sold for £12, but even more amazed when a Hoover Dustette was offered for sale! It was apparently withdrawn from sale as it was electrically faulty.

The Annual General Meeting was reported as an orderly affair. Nigel G1BUO was elected Chairman for a third year, but Ted G3DCC stood down as Secretary. He had volunteered to act as Secretary on a temporary basis when Brian G4WYG left for the Sudan in 1989. Some three years later, and approaching his 72nd birthday he believed it was time to step aside. Alan G4BWV took over the Secretarial duties. Glyn G8KKI also stood down as Treasurer, having been in post since 1977. Margaret G0ARQ, Alan's XYL, became Treasurer.

'El Presidente' passes away

Society President, Alan G3ANK, passed away on 2 May 1992. Alan, or "El Presidenti" as was his nom-de-plume, was one of the founding members of the Society. He was a staunch supporter of the Society and in the early years was a keen contest participant, often to be found at HF Field Days helping to set up the antennas, tents and supporting the operating team. HF CW was Alan's favourite mode and, being a highly accomplished operator, it was only natural that he ran several highly successful Morse classes for members. He retired to Dorset, but kept in touch with members, and always made a special effort to attend Annual General Meetings. His funeral was attended by Geoff G2CXO, Don G3KGM, Keith G3TAA, Chris G3VLT and Bob G8JNZ. The proceeds of the May surplus sale were donated to the British Heart Foundation in his honour. Alan's passing was discussed by the Board and it was agreed to preserve the name of such an important and dedicated member of the Society by continuing to show him as the Society's President in QUA, and to elect a new President at the 1993 Annual General Meeting. It was unknown by many, but Alan's widow, May, was genuinely touched and understandably proud at this gesture.

The DF Hunt was an enjoyable affair which was won by Paul G3SXE's team. Only four teams took part; two found the 'fox', located in Downe Village, some 10 miles from the Eltham start point. Although not a Cray Valley event, Dave G4NOW set up a special event station (GB2AG) from the British Vintage Wireless Museum in June.

It was announced in September that past member, Pete G3RZP, was to become the next RSGB President. The passing of Peter G0HUM was announced. The Society was well-represented at his funeral. A serving Police Officer, Peter was afforded a full Service funeral, which drew an attendance of over 450 people. The Society made a donation to a trust fund set up for his youngest daughter.

QUA changes

Derek G3XMD had printed many thousands of pages of QUA for over 25 years up to September 1992, the newsletter being printed on ancient Roneo stencil type copiers that were longer in the tooth than Derek. It was announced that both needed to be retired gracefully; the printer becoming worn and its master in need of a rest from cranking its handle every month. Predicament! How was QUA going to be printed and who would offer to do it?

Offers were sought, but no-one came forward. Measures were therefore taken to print QUA, at a cost, commercially from November. QUA printing was out-sourced to the St. Mary's Media Resource Room in Eltham High Street. As the Society was already a member of the St. Mary's Community Complex and met at one of their halls, a significant discount was achieved on printing costs. There was also a change in appearance, with a return to a coloured letterhead and a new two-column layout. Owen G4DFI arranged for the provision of letterheads and computer generated labels, while Joan, my XYL, took over the job of stapling,

The November 1992 QUA letterhead

folding, enveloping and posting QUA to those who were unable to collect their copy from mid-monthly meetings.

The Society had seen few visits by an amateur radio retailer, so the visit by Icom UK to the December meeting was particularly well-attended. The evening began with a talk about the history of the company, which had had grown from small beginnings in 1974. Chris G8GKC followed this with an impressive demonstration

88

of one of Icom's latest mobile transceivers. Some members had known Icom UK staff through the Kent Repeater Group, but we did not know at that time how close a relationship the Society was going to have with the company.

Another poor attendance at the Christmas Social led the committee to cancel future December mid-monthly meetings. This meant there would be only one Society meeting in December and one in January until further notice. Membership throughout the year remained stable at 54.

1993

Attendance at meetings in the early part of the year was poor, even though some excellent guest speakers had been invited. The Society had a long-held tradition of inviting guest speakers, but it was becoming something of a gamble to invite them in view of the low attendances. Several guest speakers were booked for later in the year, so the committee were hopeful that attendances would improve to avoid embarrassment, both to the speaker and the Society.

New Society President elected

I was honoured to be elected President of this great Society at the 1993 Annual General Meeting. My message, perhaps a rallying call to members through the pages of QUA was:

"It is with some pleasure that I write these words to you all as the Society's new President. This exalted position was afforded me at last month's Annual General Meeting for which I am indeed most honoured. I have been a member of the Society since the late sixties and served on committee first in 1969/70 under the chairmanship of Fred G3XFG. I was delighted to have been Chairman during the heady seventies (1975-78) when we could boast a membership well over 100. And we were indeed a most active Society where people were falling over themselves to become members. Our meeting place in Court Road, Eltham was definitely "standing room only". Indeed the church authorities were glad when we moved on as we broke their fire regulations at every meeting! More recently, I have been one of the Society's Vice Presidents. Having said that, I would be more than happy if Alan was still at the helm. Alas, that is not possible.

Things have taken a turn for the worse in many respects since those days but, in my role as President, I now hope – just like the sunspot cycle – that the Society again rises to a peak. I would also hope that the Society, which is still a great one, can reach the peaks of 15 years ago and once again be one of the most successful in the country. To do this we need each and every member to wipe away the apathy and do something for the Society – however small – in the next couple of years to help us get back to where we belong – at the top."

Membership, financial and limited company concerns

The new committee were concerned about the declining membership. From 54 at the end of 1992, only 37 members had renewed before the Annual General Meeting. This had implications for the year's projected cash flow. Should the committee eat into reserves? Should subscriptions be raised? The committee monitored the position at monthly meetings, but it was agreed that reserves should not be allowed to dip below £1,000, and members gave the committee a clear mandate at the Annual General Meeting not to raise subscriptions.

The annual subscription stood at £10 and some, especially the unemployed, found it difficult to pay. The committee introduced reduced or waived subscriptions in extreme cases. By the year end, membership had risen to 44, helping to improve the cash flow situation. Attendance at meetings was also a concern. With the North Kent Radio Society experiencing similar difficulties, members were encouraged to visit each other's meetings. Joint lectures were also considered. Tony G4WIF volunteered to act as publicity Secretary to try to raise the profile of the Society.

There was also 'feeling' amongst members about de-limiting the company. Advice had been taken by one committee member from a specialist in company accounting which reinforced the opinion that it was not necessary for the Society to remain a limited company. The committee discussed ways in which the company could be wound up, but the Chairman suggested such discussion was premature as he believed there would be further participation in contests.

In an effort to stir up some activity, a weekly 80m net was started on Monday evenings, and there was talk of putting on a small, but not too serious, contest station. The first 80m net attracted five members; the second, seven! Cray Valley members were starting to talk to each other on the radio again! But the enthusiasm did not last, with summertime nets limited to two or three participants.

The usually successful Surplus Sale was the next to suffer from a growing apathy. Attendance at the mid-year sale was down and there was a general reluctance of any of the 20 members present to bid for the items; and when they did, it was at ridiculous prices – one example – a pair of new 'N' elbow connectors: normal price over £8; giveaway price – 50p!

The year's Annual DF Hunt saw a unique ceremony at its conclusion; that of the 'fox' being awarded the trophy as no-one managed to find it! Located in Swanley Village, on what some assumed was an unmade road leading into some fields, four teams were left frustrated! Seeing a team with a hand portable and antenna, one member of the public stopped to enquire *"What's the problem"!*

G2MI SK
Arthur Milne G2MI passed away in October. He had been a member from 1964 to 1982 and had been the Society's President from 1965 to 1969, and a Vice President in 1969/70 and from 1974 to 1976. Arthur's numerous accomplishments, however, were in the global world of amateur radio. He was a member of the band of 'secret listeners', the Volunteer Interceptors, during World War II. In 1939 he took over the RSGB QSL Bureau, which he ran for over 40 years. He was President of the RSGB in 1954 and the main GB2RS newsreader for many years. The Society sent flowers and a card to Arthur's family expressing sincere condolences.

In comparison to the poor turnout at many meetings and the general apathy, the Christmas nets were quite well supported with 10 participants on the 2m net and seven on the 80m net.

1994

Had apathy reached a new level? The January QUA certainly seemed to suggest it with this comment from the editor – *"I'd love to meet the members* (referring to a poorly attended 'Meet the members' video evening in late 1993) *except that not many of them turn up for meetings these days."* Bit by bit Society activities were being wound down. Special event stations had been cancelled first because only a handful of members supported them, and visits were not over-patronised. Late December and early January meetings had been cancelled, followed by the Christmas Social. All for the same reason: mediocre interest. Due to poor attendance at meetings, it was even suggested that one meeting a month might be sufficient.

The Society was going through a serious phase, not helped by some speakers pulling out of talks at the last minute for various reasons, and it was noted that committee members outnumbered ordinary members at meetings. The question was asked *'So, what are we going to do about it?'* One member did write to the editor to express his views – *"God forbid that Cray Valley, so synonymous with amateur radio and all that ham radio stands for, were to close down...perish the thought."*

The Annual Construction Contest, however, saw an interesting array of exhibits. Judges Chris G0FDZ, Dave G0JBT and Paul G3SXE were hard-pressed to decide the winner. Tony G4WIF was eventually judged the winner for a neat audio filter which he had built as an add-on unit to improve his KW transceiver's CW performance. Steve G0TDJ was second for a splendidly restored Mk.7 Avo meter. He had purchased two, both non-working, at a Surplus Sale and made one working meter out of them. The judges were particularly impressed with the minute lettering on the front panel, which had been painstakingly removed and re-painted by Steve's XYL.

The Annual General Meeting was another orderly affair. 23 members attended – exactly half the Society's membership. The Treasurer reported that the Society was in good financial shape, but the Chairman once again stressed the need for members' feedback to be able to provide for their needs. It was noted that Cray Valley was not the only local radio organisation suffering apathy, with news of the winding up of the Dartford Heath DF Club.

June was very much a success. The Surplus Sale attendance was the best for a long time, with members and visitors packed into the meeting room and the amount of surplus equipment for Derek G3XMD to sell weighing down four large tables. Derek sold everything and sat back with a contented face while he watched people queueing to settle their accounts.

The Annual DF Hunt followed the Surplus Sale. Nigel G1BUO and I were again the 'fox', this time located outside the *Hare and Billet* in Blackheath. Chris G0FDZ's team won 'The Tally Ho! Cup'. One hilarious moment befell Tony G4WIF's team. Upon finding a bearing going straight through a Police van, he thought a nifty stunt had been pulled, and banged on the window and enquired if Nigel (a serving Policeman at the time) was hiding in the back! It seems their reply was courteous but to the point!

August saw a 'Meet the Members' video evening featuring the shacks of Deryck G3VLX, Chris G0FDZ, Bob G8JNZ and Paul G3SXE. Deryck's HF yagi was a particular talking point as the day before the video was taken, a ferocious storm had reduced a four element HF antenna to a two-and-a-half element! May, Alan G3ANK's widow, forwarded some of Alan's Society records for safe keeping. Bob JNZ read all the documents and said it *"...was a treasured trip down memory lane"*. It is unfortunate that the whereabouts of this material is unknown, and cannot be included in this history of the Society or Chris G0FDZ's 'Project 70' memorabilia archive.

With no Society mid-monthly meeting arranged for December, Chairman Nigel G1BUO booked a table at an Indian restaurant in Bexley for an informal meeting of a few committee members disappointed that no Christmas Social had been arranged. The event was mentioned in QUA in case others wanted to attend and the evening was a huge success with even traditional English-grub-only diehards apparently converted to the manager's mild 'beginner' curry.

1995

The year began with a well-attended donated surplus sale and was followed by a *'trouncing'* of the Darenth Valley Radio Society in an inter-society quiz. There were five entries to the Annual Construction Contest, the minimum requirement for a contest to be held. Tony G4WIF won the Reigate Cup for a second consecutive year with a spectrum analyser adapter. It was reported in QUA that Steve G0TDJ was a close contender for the top prize with a Direct Conversion receiver.

Members at the Annual General Meeting voted by a majority to ban smoking at all times in the meeting room. Nigel G1BUO was elected as Chairman for a sixth year, but a lack of nominations meant there was no Vice Chairman and only two committee members; Alan G4BWV and Margaret G0ARQ continued as Secretary and Treasurer respectively. A suggestion was made that the Society examined in depth its current status and its plan for the future. A sub-committee of Chris G0FDZ, Bob G8JNZ, Owen G4DFI and Paul G3SXE was formed to undertake this major study.

In response to members' wishes, a 2m Society net was introduced in June to try to increase activity, but it was poorly supported and disappeared by the end of the year.

Owen G4DFI and Paul G3SXE were the 'fox' for the Annual DF Hunt. They were located at Hodsall Street, near Vigo Village. Only three teams took part and the 'fox' was not found. The 'fox' therefore won 'The Tally Ho! Cup' again.

The summer surplus sale saw an audience of 21, but seven were visitors. There were only four vendors and Derek G3XMD had a real job in overcoming the general apathy of the audience. However, he did succeed in selling most of the items and made a reasonable profit for the Society.

QUA received a new look in September. Well, just the front page! The famous coloured logo disappeared. Over the years Owen G4DFI had arranged with a work colleague to print copies of the letterhead front page at no cost to the Society. However, his colleague retired and the stock of letterheads had been exhausted. As QUA was one of the more costly items of Society expenditure, obtaining a professionally produced letterhead in traditional red would have significantly raised expenditure, so Bob G8JNZ and Tony G4WIF designed a number of computer generated letterheads from which they chose the one they liked best.

GX3RCV was aired from the Eltham meeting place at the main August Society meeting. 19 members were present. Activity was on HF, with Owen using his Yaesu FT-990 transceiver into the Society's 5-band vertical. An SWL station was also running – Steve G0TDJ's FRG7700 into Ted G0ULL's home-brew loop antenna and a packet radio station was also on show, which attracted interest from several members.

In October, the Darenth Valley Radio Society were successful in the return inter-society quiz. Ten members made the trip and our team of Ted G3DCC, Owen G4DFI and Alan G4BWV were leading but were undone by a series of questions about commercial satellites, frequencies and their modes of operation. A final 'TV commercials' round, which saw Ian G7PHD replace Alan G4BWV, failed to claw back the deficit, and the Cray Valley team lost a very enjoyable contest by 90 points to 70.

The sub-committee examining the Society's status and plan for the future had met regularly during the year and devised a questionnaire. They planned to ask each member the questions individually, with the responses remaining anonymous. It was hoped that the exercise would obtain the best possible overall member representation about the future of the Society. A report to the committee was planned for 1996.

After the success of the informal Christmas get together in 1994, Nigel G1BUO arranged a return visit to the same venue, again with a good turn-out. The year ended with only 40 members, the lowest Society membership since 1963.

1996

Although not a Cray Valley event, it seems right to record that Chris G0FDZ won the first RSGB 'Microwave Award' for his home-brew 24GHz transceiver. This was a truly fitting reward for Chris' dedication and perseverance in building home-brew equipment over many years.

The 1996 Annual Construction Contest was attended by only 20 members, but there were 20 entries varying from a 25 volt twin power supply, through to yagi antennas, amateur band receivers and Baycom and Hamcom modems. Chris G0FDZ and Dave G0JBT awarded the Reigate Cup to Ted G0ULL for a home constructed power supply, second was a 2m yagi designed and constructed by Steve G0TDJ. Certificates of merit were awarded to three short wave listeners for their home constructed amateur band receivers.

Alan G4BWV began researching the Society's membership, officers and committee from 1946 to 1995 and asked members to contact him with any information. The support he received was such that he was able to compile an excellent record, although he was unable to trace any officer and committee details between 1953 and 1961. The information he obtained has been extremely valuable in compiling this history of the Society.

The Annual General Meeting was the 19th annual meeting of the limited company, but there were only six nominations for the Board. All were proposed and approved en bloc. Some time was spent discussing the direction of the Society and how to gain new members, especially younger members, with a suggestion that the Society became more involved in the amateur radio Novice Scheme. Looking forward, Derek G3XMD suggested that as Greenwich Council had won the franchise for the Millennium celebrations, the Society should consider putting on an extra-special event station. See how that idea unfolded in a few pages time!

'Status and future' sub-committee report presented

Just after the Annual General Meeting, the sub-committee examining the Society's status and future presented its extensive report to the committee. The telephone-completed questionnaire enabled members to state their opinion in confidence; some very clear views were expressed. The sub-committee summarised their findings and concluded that the committee had generally been *getting it right'* and that the membership was mostly happy with the way the Society was being run. The questionnaire had elicited responses that showed the committee was providing a service but in a one-way direction, i.e. assuming the membership's requirement without asking them, such that it could not be proved whether the activities the Society had been providing were actually what the members wanted.

The answers to certain questions were so surprising as to indicate a nucleus of very active radio amateurs within the Society who were not only on the air regularly but who were willing to support special event and contest stations. After the lean activity witnessed through the '90s, this finding suggested a few less than truthful answers!

In general terms, members did not seem unduly bothered about the Society's limited company status, but there was a feeling amongst some members that the Society had no real need to remain a limited company. As the matter was soon to come to a head, it is worth spending time here to consider the stated pros and cons of the Society withdrawing from its limited company status:

<u>Pros</u>

- the current auditor would be unable to audit the accounts indefinitely. De-limiting would enable the accounts to be audited internally
- engaging a chartered accountant in place of the current auditor would likely be at a greatly increased cost
- the lengthy and cumbersome planning occupying half of the committee's time would be lessened if the Society were not constrained by company law
- the Society no longer participated in external activities and was unlikely to do so unless there was a surprising upsurge in activity.

<u>Cons</u>

- insecurity about the prospect of de-limiting possibly leading to members leaving the Society
- the increased inflationary costs if the Society decided, at a future date, to re-incorporate
- separate third party insurance would be required.

After balancing the opposing views, the sub-committee's recommendation came down on the side of the Society remaining a limited company, with a further recommendation that ways should be found to minimise the impact on both committee and ordinary members.

Overall, the sub-committee recommended that the 30+ proposals in their report be implemented over a two year period. It was further recommended that the committee create a quarterly agenda item to review implementation of the proposals.

Moving back to Society meetings and activities, the May Surplus Sale attracted 30 people. Although many of the items were not sold, the attraction at sales in those days was the never ending supply of John G3LNT's 'Bumper Bundles' – bags crammed full with surplus components, plugs, connectors, etc. – and the banter between auctioneer, Derek G3XMD, and the audience, which always made for a thoroughly entertaining evening.

By comparison, the DF Hunt was poorly supported. Only two teams took part. The 'fox' (Owen G4DFI and Paul G3SXE) were located next to the Darenth Valley Radio Society's meeting place in Crockenhill. The winning team of Nigel G1BUO, Steve G6SDO and Ian G7PHD received 'The Tally Ho! Cup'.

After an absence of organised visits for a few years, return trips were made to the British Vintage Wireless Museum and Littlebrook Power Station during the summer. Attendance at both could have been better, but those who attended were treated to two first class tours. A quiz with the Darenth Valley Radio Society was also arranged. A very enjoyable evening was made even better by a 64-50 success.

50th anniversary celebrated

October 1996 saw the Society's 50th anniversary month. Three meetings were arranged, with the highlight being a well-attended anniversary party. As well as a 'normal' QUA, a QUA anniversary extra and a special QUA anniversary supplement were issued. The anniversary supplement front cover featured the agonised creature 'QRM' that adorned the Cray Valley Amateur Transmitting Society's well-produced magazine for some years from 1947. The supplement was a 10-page potted history of the Society up to that point. It also reproduced extracts from previous anniversary QUAs.

G8LDV, unknown, G8JAD, G8DYN, G4IPZ, G3DCC, G3YJW, G0FDZ and G8GGP

A 'QUA Extra' provided details of the month's activities. The first meeting was a well-attended surplus sale, the second was an RSGB video depicting 50 years of amateur radio, and the third was an excellent celebratory evening. 50 members,

98

past members and guests attended. It was a wonderful occasion, especially with so many well-known past members in attendance. Eight Chairmen, past and present, attended – Nigel G1BUO, Alan G4BWV, Deryck G3VLX, Martin G3YWO, Bernard G8LDV, Owen G4DFI and me. The evening featured a static display of memorabilia, plus a re-run of the 40th anniversary video, and a cake-cutting ceremony.

The month also featured a successful Golden anniversary award for working members, the Society callsigns were aired throughout the month, and several activity periods encouraged members on the air. 21 claims for the anniversary award were received. QUA noted that the three meetings saw a total attendance of 106.

The Gold Award

1997

The year began with news of the passing of George G2DHV who had been a member at various times since 1949, and Fred G3XFG who joined the Society in 1966 and had been Chairman between 1969 and 1971. Fred, of course, was also a key figure in the Society becoming a limited company.

The Construction Contest saw 12 entries. Chris G0FDZ and Dave G0JBT again adjudicated the competition and awarded the Reigate Cup to Peter G0GIR for a home-constructed infra-red remote control repeater. Second place went to the 1996 winner, Ted G0ULL, for a QRP transceiver. Third place went to Tony G4WIF for a QRP SWR/Power meter.

The June DF Hunt was unfortunately affected by unseasonable weather which deterred many members from taking part. Owen G4DFI and I were the 'fox' together with my two young children Clare and Simon. We were located in the car park of a public house in Westwood, near Bean. For the second time in four years, the 'fox' won 'The Tally Ho! Cup' by default. Clare and Simon were delighted to pick up another trophy, following the presentation to them of the *Newcomers' Trophy* at the April Annual General Meeting for their work in folding, stapling and enveloping QUA following the untimely passing of Joan, my XYL, during the previous year.

The July Surplus Sale was quite well attended. Derek G3XMD auctioneered in his inimitable style and after a slow start, the G3LNT 'Bumper Bundles' got the evening moving. This was the first surplus sale I had taken Simon to as, at the age of nine, he was showing an interest in following my footsteps into the hobby. QUA remarked that a notable treat for the audience was Simon bidding for some surplus equipment and that I had to pick up the substantial tab!

The August QUA introduced yet another new look front page header. Designed by Tony G4WIF, elements of the header are still in use

The August 1997 QUA letterhead

today on letter headings, event posters, QSL cards, and even on the front of this book!

Notice was given by the Secretary that an Extraordinary General Meeting was to be held in September to discuss and vote on a resolution 'To de-limit the Cray Valley Radio Society Limited'.

A return to Society status

As President, and Chairman of the Society when the decision was taken to become a limited company, I began the debate about the Society reverting from a radio society limited by guarantee to a simple radio society. In 1977, members of the Society decided it was right to become a limited company. Moving forward 20 years and with contest activity and special event stations in the distant past, members were of the opinion that the extra protection limited company status brought was no longer required.

The committee's statement of reasons gave two other reasons why they believed it to be the right time to dispense with the limited company status: the Society's auditor was giving up the auditing side of his practice, and members were reluctant to stand for office while the Society remained a limited company. The committee expanded on both reasons. The auditor had provided a quality service at a very competitive set rate of £100 for many years. Approaches to other auditors to provide an audit for a similar price had proved impossible: a minimum charge of £300 had been the best quote. The committee believed that employing a new auditor for such a fee would require a major increase in the level of subscriptions – from £10 to £17 for full membership in the first year alone, and with the prospect of further year-on-year increases in both audit fee and subscription levels. The committee was also concerned that the Society's excellent financial position would quickly be eroded if limited company status was retained and a new auditor appointed.

The second reason was that it had become apparent for a few years that members were reluctant to stand for office while the Society remained a limited company. With the Chairman, Secretary and Treasurer all announcing that they would not be seeking re-election at the 1988 Annual General Meeting if limited company status was retained, there were very real implications for the Society's continued existence.

Three resolutions were presented to the Extraordinary General Meeting –

The Members of Cray Valley Radio Society Limited in attendance at this meeting hereby Resolve that verbal proxy votes as well as written proxy votes may be counted in any vote for or against the following Resolutions;
Members of Cray Valley Radio Society Limited hereby Resolve that the Company shall cease to exist on 31 December 1977 and Cray Valley Radio Society return to Society Status;

Following the above move, Cray Valley Radio Society Ltd., Members hereby resolve that all assets, finances and monies be passed to Cray Valley Radio Society as from 1 January 1998.

Members voted in favour of the resolutions. Companies House informed the Secretary in May 1998 that Cray Valley Radio Society Limited was defunct.

1998

Few back copies of QUA were available to provide a more detailed look at activities during 1998. However, it is fascinating to report that only a few months after the Society ceased being a limited company, it embarked on the biggest special event station in the British Isles since GB2LO (the RSGB exhibition station established in association with *'The Daily Mirror'* in 1968 to celebrate the *'City of London Festival'*). The station was to be an enormous success, and re-launched Cray Valley Radio Society as a global name in the amateur radio world. Planning for the event took 20 months and involved a 13-man Cray Valley sub-committee, which I led.

With concerns about standing for committee whilst the Society held limited company status resolved, the Chairman, Secretary and Treasurer were all voted back into their officer positions by the membership at the Annual General Meeting. The three committee members elected were Paul G7RCE, Richard G8ITB and Bob G8JNZ.

Early planning for the Millennium – 'Project ECHO'

The suggestion of celebrating the new Millennium was first raised by Derek G3XMD at the 1996 Annual General Meeting. From that small seed, the idea was floated at the 1996/97 St. Mary's Community Complex Annual General Meeting. The idea was well received and subsequently approved by Len Duvall, Leader of Greenwich Council. 'In principle' approval was given by the St. Mary's Management committee, and a joint working group, comprising St. Mary's and Cray Valley, was established with a first meeting held in May, and funding opportunities researched. A budget of around £16,000 was identified. In November, Len Duvall introduced the Society to Sir Jocelyn Stevens, Chair of English Heritage, with a view to using Eltham Palace as the venue for the special event station. Sir Jocelyn was approached by letter. He replied in December having passed the proposal to his Deputy Director for London to discuss it with Society representatives in more detail. A meeting was arranged at English Heritage's offices in London for 20 January 1999.

1999

Although a full meeting programme was arranged, including a poorly supported Construction Contest, QUA was devoid of any other articles during the year to suggest that the Society had been involved in any meaningful external activity. One reason for this was likely to have been because so many members were involved, in one way or another, with preparations for the Millennium special event station.

'Project ECHO' planning

Why 'Project ECHO'? The team wanted a punchy name which was closely associated with their aim of contacting as many radio amateurs around the world as possible from Eltham. The 'ECHO' acronym stood for 'Eltham Contacting Hams Overseas'.

The stated aims were to:

- *make contact with radio amateurs around the world as they crossed into the new Millennium*
- *provide a top-of-the-range operating environment capable of providing 20,000 amateurs worldwide with the opportunity of contacting the Millennium borough, and*
- *improve the lifetime learning of schoolchildren by providing radio and electronics tuition, allowing them to see the station in operation and pass 'Greetings Messages', visit the static and inter-active radio display, and have the opportunity of purchasing a receiver kit to construct as a school or class project.*

The organising committee worked tirelessly throughout the year to plan the biggest special event station the British Isles had seen for over 30 years. Following a meeting at English Heritage's London offices in January, agreement was reached that the special event station could be located at Ranger's House in Greenwich Park between 31 December 1999 and 29 February 2000.

Ranger's House was a particularly suitable location for the project as it was situated directly on the Meridian Line. Additionally, until 1752, the New Year in England began on 25 March. England followed the ancient Julian Calendar, which lagged 11 days behind the rest of Europe which followed the Gregorian Calendar (introduced by Pope Gregory XIII in 1582). It was Lord Chesterfield, owner of Ranger's House, who introduced the Calendar Act of 1752, replacing the Julian Calendar with the Gregorian Calendar and bringing England into line with the rest of Europe. At the same time, Lord Chesterfield pushed through a law making 1 January New Year's Day instead of 25 March.

Society members were excited about the opportunities afforded by operating from a prestigious location which commanded a truly excellent geographical position, and hoped the project would catapult the name of Cray Valley Radio Society around the world.

A number of factors accounted for a successful planning phase. These included:

- approval to operate from the Greenwich Meridian
- obtaining a very special, special event callsign – M2000A
- support from English Heritage and the London Borough of Greenwich
- the award of a total of £7,000 in Grant Aid from the Greenwich Celebration Fund and the Millennium Experience Awards for All scheme
- securing 'in kind' support of almost £100,000
- the granting of planning permission for three mobile masts.

The project received wide publicity at home and abroad. A tri-fold colour leaflet was issued and a dedicated website established – www.qsl.net/m2000a. Some of the organising team were even interviewed live on 'The Giles Brandreth Show' on LBC radio.

An arrangement was made with English Heritage for any Society member to enjoy free entry to Ranger's House at all times. To cater for those radio amateurs who wanted to help with the set-up or operate the stations, a new class of Society 'temporary membership' was created. For a small fee, around 60 'temporary members' joined the Society; everyone benefited, except English Heritage - it brought welcome revenue to the Society, some high profile amateurs were recruited to be part of the team, and they did not have to pay the daily rate to gain access to Ranger's House.

Millennium skeds were arranged with 55 DXCC entities at their midnight, and a special launch function arranged. An M2000A award scheme was devised, and a special 4-sided QSL card designed and printed.

Station objectives and configuration: At HF, three stations were built (HF1, HF2 and HF3) that were capable of reliable communication on the nine amateur HF bands (1.8 – 30MHz), with high contact rates on up to three bands simultaneously. On VHF, two stations were built (VHF1 and VHF2) that were capable of reliable local and long distance communication. A separate packet datacomms station was also built which provided a local Bulletin Board and access to the DX Cluster at

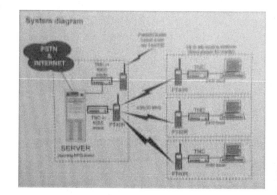

435MHz. Logging was by John G3WGV's Turbolog program. It formed part of the packet radio-based network and was linked to a server located away from the five operating stations. Contact details (QSOs) were sent automatically to the server as they were made. QSO details were uploaded to the M2000A website daily. John used M2000A to further develop the Turbolog program. A diagram of the 435MHz system is shown on the previous page.

Antenna party: Monday 20 December 1999 dawned cold and frosty. Temperatures of -6C were registered as 10 Cray Valley members made their way to Ranger's House before sunrise to begin the major job of assembling seven yagi antennas for HF and VHF and erecting three M80 mobile towers loaned by South Midlands Communications (SMC), Strumech Versatower and Chris G3VHB. SMC also sponsored the event by providing five antennas – Create AFA40 7MHz 2-element yagi; Create CY103 28MHz 3-element yagi; Cushcraft 154CD 21MHz 4-element yagi; QTEK211E 11-element 2 metre yagi; and a QTEK65E 5-element 6m yagi. Other antennas were provided by Dave G4BUO (Hygain TH5)

and Neville G3NUG (Cushcraft A3WS with 10MHz extension). All the antennas had been partially pre-assembled to save time. Although the antenna and tower erection went smoothly, it took two days to complete the build due to snow and the shorter late December days. The guying arrangement for the 70' tower located on a tarmac area was particularly impressive. Through the good offices of a local crane company, three 2-ton steel blocks were manoeuvred into position by the crane to provide the guying positions for the TH5 and CY103 yagis. All the antenna work was closely supervised by HighLine Communications Ltd. Coaxial and rotator cable was provided by Mike G3SED and Neville G3NUG. Once the antennas were in place and the towers raised, 500m of cable was routed into the M2000A shack.

Assembling the M2000A station: All the inside work to assemble five stations and set up the display material was completed over three days (28 – 30 December

1999). Cables were run, patch leads soldered, and equipment tested. Fool-proof switching was installed to avoid inter-station interference and to ensure switching antennas between stations was kept to a minimum. All the transmitting equipment was provided by Icom UK: IC-756PRO and IC-PW1 linear amplifier, 2 x IC-746, IC-775DSP and an IC706MkII; the dawn of a rich relationship between the company and the Society.

The exhibition: While the radio station was being assembled, Ian G7PHD and Steve G0TDJ and others established the static and 'hands on' exhibition charting amateur radio from Marconi to the new Millennium. Display Colour Ltd provided 12 3' x 2' display boards, and the RSGB and the British Vintage Wireless Museum provided radio exhibits from the early 1900s. The contrast between 1900 and 2000 was demonstrated by a display of the latest semi-conductors and micro-processors side-by-side; the latest Geoclock software and the Icom IC-PCR1000 computer-driven communications receiver. There was also an RSGB bookstall.

The launch: Lord Brian Rix G2DQU launched the event at 10.30z on 31 December 1999. The Mayor and Deputy Mayor of Greenwich, Hilary G4JKS (1999 RSGB President), Don G3OZF (2000 RSGB President), Mark 2E0APH (British 'Young Amateur of the Year'), plus representatives from English Heritage, Icom UK, SMC and Cray Valley Radio Society were present. A champagne reception was enjoyed, courtesy of J Sainsbury PLC, and a special celebration cake, supplied by Marcia 2E1DAY, was cut by Lord Rix.

The first contact between M2000A and ZL6A at midnight New Zealand time was conducted by Lord Rix G2DQU with several 'Greetings Messages' exchanged.

Once the sked was complete, all five stations at M2000A became active and the pile-ups began! In the first five hours of activity 1,176 QSOs were in the log, and a number of other first day skeds completed, including one with 9G0ARS in Tema in Ghana. This sked had been specially arranged so the Mayor of Greenwich could pass a 'Greetings Message' to officials in Tema – the only other city to the south of the British Isles located on the 0° Meridian. By midnight, 2,600 QSOs were in the log.

...and so Cray Valley Radio Society and its members and temporary members, now totalling over 100, embarked on two months of activity which put the Society's name in lights.

Lord Rix G2DQU making the first contact with ZL6A

Chapter 7: Out of the doldrums...turning on the activity tap: the 'noughties'

This chapter moves away from the mundane 90s to the activity-charged noughties. Looking back, losing limited company status breathed new life into the Society. Gone was the apathy and the reluctance to become involved in external activity. The change was so very noticeable. These next pages are full of the best of Cray Valley Radio Society: special events, contest success, a growing membership and increased attendances at meetings. The future was rosy, starting with the Society's first ever major special event station – the wonderful times enjoyed at M2000A.

2000 and M2000A

The main event in 2000 was the M2000A special event station at Ranger's House. Although the English Heritage property was not open to the public on New Year's Eve or New Year's Day, the operators completed a number of pre-arranged skeds with other Millennium special event stations as their countries entered the new Millennium. By the end of the operation, M2000A had contacted 52 special Millennium stations. Although a faulty fire alarm curtailed activities temporarily (and brought out the Fire Brigade!) on New Year's Day, the station was on the air daily between 08 - 19z until 29 February 2000. On most days, two or three of the five stations were active but contact totals varied depending on the number of operators and the number of visitors. At weekends, contacts generally peaked between 1,500 and 2,250.

One of the specific aims of the project was to allow students of all ages to learn about amateur radio and pass a 'greetings message' over the air. A timetable of visits was arranged which saw 15 schools visit 'Project ECHO'. During one visit all 33 students passed a message, keeping the operator busy for over an hour! Each student who visited received an RSGB 'goody bag', kindly sponsored by Icom UK.

Over 200 licensed amateurs and short wave listeners visited M2000A. It was particularly pleasing to welcome a number of DX amateurs, including some from Japan, America and Europe. The team's best 'eyeball' DX was WH2M from Guam.

M2000A achievements: The event was a huge success. Cray Valley members made 47,752 QSOs in 202 DXCC entities in only 567 hours of operation (equivalent to only 23 days round-the-clock operation). This was a record number of QSOs made by a special event station at that time.

First and foremost, over 70 Cray Valley Radio Society members activated the most prestigious and highest profile amateur radio station in the United Kingdom since GB2LO in 1968. The success was reported on radio, television and in the local, national and international press. The team received a number of compliments from home and abroad for achievements in respect of a) amateur radio in general; b) amateur radio in the British Isles; c) the consistent strength and quality of signals; and d) the excellent skill of the operators.

Dave G4BUO with a group of young students

M2000A attracted 1,200 visitors, many experiencing amateur radio for the first time. The team demonstrated amateur radio to almost 400 students, many of whom passed 'greetings messages'. Apart from the high contact numbers and the number of different countries contacted, M2000A achieved 'Worked All Zones', 'Worked All States' and contacted 100 different Island groups, as well as other national and local awards. 19,256 CW contacts were made, with 28,157 on SSB and 339 on FM. The contact rate over the whole period exceeded 84 contacts per hour. It is interesting that M2000A contacted more stations in Argentina than in Ireland, more stations in Korea than in Northern Ireland, and more stations in Mongolia than in Jersey!

There was occasional contest activity from M2000A. Pride of place went to the team of G3WGV, M0CFF, G3XWK, G4VXE and G0OPB who made 1,694 contacts in the ARRL DX CW contest.

Privileges: M2000A was granted a number of licensing privileges by the Radiocommunications Agency (RA). Class B and Novice licensees could, of course, operate as long as a full licensee was present but there was a special dispensation which allowed non-licensed people to conduct QSOs under supervision and, with the station open to the public, for 'greetings messages' to be passed by visitors. Permission was also given to obtain agreement from a number of overseas RA equivalents to allow third party 'greetings message' traffic with M2000A by their radio amateurs.

A lasting memory: The final contacts at M2000A were made on 29 February 2000 but the event lives long in the memories of those who were involved. A 4-sided QSL card has been sent by QSL Manager, Owen G4DFI, to confirm over 35,000 QSL requests, with cards even being received in 2016! Over 300 operating awards were issued by Awards Manager, Clare RS102891. A mountain of video footage, audio tape and stills was edited into a full length video by Paul G3SXE, with 80 copies sold. A stand was established at the RSGB HF and IOTA Convention later in 2000, and 12 lectures given to UK amateur radio societies during 2000 and 2001. English Heritage, Greenwich Council

Clare RS102891, Simon RS177448 (now 2E0CVN) and Bernard G8LDV (now G3NPS) at the RSGB and IOTA Convention

and Icom UK issued major press releases, and finally, Greenwich Council honoured the team with a reception at Woolwich Town Hall to recognise the publicity that the event had afforded the Borough.

A look at the rest of 2000

The year started with great promise. Swelled by those who had joined the Society to be part of M2000A, membership rose to 108. But it has to be said that although M2000A was not everyone's 'cup of tea', it projected the Society into the forefront of the amateur radio world. There was hope that the event would enhance the Society and its activities. Bernard G8LDV was elected Chairman at the Annual General Meeting and a new committee comprising Andrew M0BXT, Conrad G6SDO, Richard G7GLW and Bob G8JNZ was formed.

The Annual DF Hunt was better supported than in recent years. Seven teams (14 members) departed Eltham seeking to find the 'fox', which was located at North Cray in the guise of Owen G4DFI and Richard G8ITB. Dave G3JJZ and Pat G0BRV were one of only two teams to find the hidden station in their first DF Hunt. They received 'The Tally Ho! Cup' at the post-DF Hunt Social, which attracted 25 members.

The Society held an extremely successful Summer Social at Richard G8ITB's QTH in July. Great weather, excellent traditional fayre and plentiful liquid refreshment was enjoyed by over 20 members.

Summer Social 2000: L-R G6SDO, G8LDV, G0FDZ, G8JNZ, G7GLW, XYL G0WLF, G0WLF, G4DFI, G3GJW, G1BUO, BRS32525, RS102891, XYL G8LDV, XYL G1BUO

The Society saw some excellent and diverse lectures during the year, including talks by Don G3OZF (now G3BJ) on taking the RSGB into the new Millennium, Paul G3SXE on marine communication, Richard G7GLW on Routemaster preservation, Colin G8XDR on St. John's Ambulance communications, Chris G0FDZ on satellite communication, and Nobby G0VJG on his DXpedition to the island of Sark.

2001

The Society felt the benefit of M2000A in 2001 with activity very much back on the menu. The Society also gained a new website manager, Andrew M0BXT, a new website, www.cvrs.org, and new QUA editors, Clare RS102891 and Simon RS177448. Bob G8JNZ 'retired' as QUA editor having been involved in its production for over 20 years. Bob set high standards for the newsletter, having seen it obtain 4th place in a *Practical Wireless* national newsletter competition. Each month, issues were always of excellent quality, produced to time, and kept members informed and entertained.

The year's first activity saw the Society asked to take over the organisation and running of the popular White Rose LF Bands SWL contest. Nine entries were received, including a winning Multi-operator entry from 'The London SWL contest team', made up of Cray Valley SWLs.

Derek M0BGX arranged a leisurely tour of *HMS Belfast* moored in the Pool of London, with an opportunity for members to operate the GB2RN station and

Nobby G0VJG and John G3XWK at GB2RN

enjoy a complimentary buffet. Each member had his own highlights, but the best memory recorded in QUA was one of sitting on the captain's chair on the Compass Platform looking out over the River Thames, listening to a commentary of the Battle of North Cape on 26 December 1943 which ended with the sinking of the German battle-cruiser *Scharnhorst*. It was suggested that with a good imagination, members could actually imagine the *battle* raging in front of them!

Bernard G8LDV was re-elected Chairman at the Annual General Meeting, with Richard G8ITB remaining as Secretary. Richard G7GLW, Dave M0BGR, Steve G6SDO and Wilf G0WLF were elected as the year's committee, and Andrew M0BXT was co-opted to the committee as webmaster.

16 members made 89 contacts with the Five Star DX Association's (FSDXA) D68C DXpedition to the Comoro Islands. A club competition was organised by FSDXA to stimulate activity with the DXpedition: Cray Valley was placed 7th. The M2000A team received a certificate of appreciation from the FSDXA for their donation to the DXpedition.

The annual Construction Contest saw only five entries. The winner of the Reigate Cup was Dave G0JBT for a 28/70MHz transverter system based on the *Practical Wireless 'Meon'* system for the transverter and the *Cirkit* 50/70MHz power amplifier unit. A new 80m net, with Owen G4DFI as net controller began. Unfortunately, planned activity in VHF NFD and the IOTA contests had to be cancelled due to the outbreak of Foot and Mouth Disease in the UK. Four teams set off for the Annual DF Hunt, and for the second year, Dave G3JJZ and Pat G0BRV won 'The Tally Ho! Cup'. None of the other 'hounds' found the 'fox', which was located near Dartford Heath.

New found special event enthusiasm

With new found enthusiasm, three special event stations were arranged. They were GB2SJS, celebrating the centenary of the Nave at St. John's Church in Sidcup, GB2FB for the British Wireless for the Blind's *'Transmission 2001'* and GB8ST at Clare and Simon's Crown Woods school for their Science and Technology Festival - *'SciTech 2001'*. GB2SJS was on the air for eight hours and made 305 contacts, largely due to some excellent 40m conditions, which enabled Richard G7GLW to make 257 contacts. A Sidcup Scout troop visited the station and each of them passed a 'greetings message'. A 'Learn Morse in 15 minutes' training course was also very successful. The event obtained publicity in the local press.

'Transmission' was a national fund-raising event organised by the British Wireless for the Blind (BWBF) to support their work to provide specially designed audio equipment, free for life, to UK registered blind people. Organised by Nobby

G0VJG, 15 members manned three stations and made 1,988 contacts with radio amateurs in 90 different countries in 30 hours. The event raised £2,167 in sponsorship, which won the Society a 7m Tennamast for the highest number of contacts and the most money raised on behalf of the BWBF. Once again, the team achieved good publicity – both in the press and on two different commercial broadcast radio stations.

'SciTech 2001' was another well-supported event. Dave M0BGR and I were both on the Crown Woods School Parent/Teacher's Association organising committee and inputted heavily to the arrangements. 12 members and three students helped erect the antennas in advance, and 15 members arrived on the day to make 246 contacts. 'Learn Morse in 15 minutes' training courses were extremely popular amongst the students, such that extra courses had to be added! Visiting Cub Scout leaders saw the demonstration and this resulted in the Society giving two separate presentations at a later date, including Morse training, to 30 Cubs and Scouts. Richard G8ITB's 'Introduction to amateur radio' talk was also well received, and the RSGB set up a popular 'Travelling Bookstall'. Arrangements were also made for a Charlton Athletic FC presence at the event, with 'greetings' passed by one of the first team players.

55th anniversary

30 members and guests attended the Society's 55th anniversary party, organised by Wilf G0WLF. Roger G3MEH, the RSGB's Regional Manager, was the guest speaker. A static display of memorabilia was on show, and Nigel G1BUO arranged a superb buffet.

Members at the 55th Anniversary evening

The October RadCom made nine references to the Society, including an article about Clare and Simon taking over the editorship of QUA. George Brown, Technical Editor of RadCom, wrote *"...QUA is the best Society magazine we get here at HQ, so keep it up"*. A traditional Christmas Social was back on the agenda, and was supported by 21 members and guests.

2002

2002 was notable for a second special-special event station, this time from Windsor Castle to celebrate HM The Queen's Golden Jubilee. Much more on that shortly.

But first, the Society received a certificate early in the year confirming DXCC for M2000A. Also in January, the RSGB 2m AFS contest was entered for the first time. G3JJZ, G4BUO, G4DFI, G8ITB and G0WLF took part and achieved a decent placing, being placed 6th. Later in January, nine members formed three teams to enter the RSGB's 80m AFS SSB contest. The 'A' team of Dave G4BUO, Richard G7GLW operating G3RCV from Smudge G3GJW's QTH, and Tim G4DBL made 713 contacts; G3JJZ, G4DFI, G0VJG, G0FDZ, G0WLF and G3SXE were the other entrants.

The activity theme continued through the year with an entry into the CQWPX SSB contest, VHF NFD, a second Wireless for the Blind station and another event at Crown Woods School.

Successful contesting return
Following the upsurge in activity, the committee believed it was time to venture into big-time contesting again, and the Society successfully applied for a new, short, contest callsign – M8C. It was used for the first time in the CQ WPX SSB contest from the 9th Dartford Scouts HQ in Dartford. Nine members helped to erect the antennas, the most troublesome being a 40m delta loop. Dave G4BUO had constructed and tested it at his QTH, but it was not resonant in band at the Dartford QTH. So, reverting to Plan B, an inverted-V dipole was used!

For a first serious contest, the team made 2,211 contacts with 807 prefix multipliers for a score of almost 4M points, gaining 2nd place in England.

Simon M3CVN (now 2E0CVN) and Richard G7GLW operating M8C in CQWPX

Interestingly, the team found that 'Mike Eight Charlie' worked well when working the USA, but 'Mexico Eight Canada' seemed better suited for the Europeans. One witty American asked if *'...Mike was OK after eating Charlie'*!

Thanks to Clare RS102891's ingenuity, her 'MateCitchen' (M8C kitchen) provided the team with hot and cold drinks, cakes and biscuits throughout the weekend, which made £10 for Society funds!

With the launch of the new Foundation Licence on 1 January 2002, the committee believed the Society should consider organising training courses. Dave M0BGR was tasked with undertaking a feasibility study to look into member support, and to report back to the committee. Logistics and Insufficient support meant it would be a few more years before this initiative was realised.

The Annual General Meeting saw Bernard G8LDV re-elected as Chairman and Treasurer. In his address, he reflected on the upsurge in external activity and that it had been achieved with a membership one-third the size of the Society's peak membership in the mid-1970s.

GB50 – Windsor Castle

Although Dave G4BUO, Chris G0FDZ and I said *"Never again"* after M2000A, celebrating HM The Queen's Golden Jubilee was just too good an opportunity to miss! So I got the core team together again, with Dave G4BUO as 'HF Manager' and Chris G0FDZ as 'VHF Manager'. Organising any sort of special event station is quite a task, but with the special considerations of the Castle (access, security, appearance), the job was huge. However, with the invaluable experience gained at M2000A the event was planned in only six months, a third of the time taken to organise the Millennium station. A number of site visits were required to agree a suitable location for the event and to conduct EMC tests, including one to confirm that GB50 would not cause TVI to the Queen's television!

Following on from the unique M2000A callsign used for the Millennium station, the RA issued a second unique callsign – GB50 (Golf Bravo Fifty) – for use at Windsor Castle. The station was the main amateur radio operating event in the British Isles to celebrate Her Majesty's Golden Jubilee. It was located in a specially erected 50' x 20' marquee situated on the North Terrace of the Castle. The station

The marquee at GB50 with Steve G4JVG (now PJ4DX)

was operational from 7am – 10pm daily from Wednesday 29 May to Sunday 9 June 2002. In addition to the GB50 station, the marquee featured 'The Amateur Radio Experience', open to the public daily between 10am and 5pm. The RSGB, in conjunction with the Burnham Beeches Radio Society, organised that part of the event. The objectives of the event were twofold – to raise awareness of amateur radio, and to build a station that was capable of reliable communication with radio amateurs around the world on eight amateur HF bands (3.5 – 28MHz); and on the 50MHz and 144MHz VHF bands. All the team members had to be security vetted by the Castle security services and had to wear ID lanyards at all times. A dedicated website was also established.

Station design and set-up: The station design was closely based on the successful configuration used at M2000A. Once again, there were three HF stations (HF1, HF2 and HF3) and two VHF stations (VHF1 and VHF2). As the event took place in the summer, conditions were very different from M2000A with QRN and daytime absorption on the HF bands, but at VHF, Sporadic-E conditions proved to be excellent, allowing some exceptional DX contacts.

Planning the operation was meticulously scrutinised and pre-approved by the Castle authorities. With agreement reached, the Cray Valley team were able to position one trailer tower and assemble the yagi antennas, while *Marquees Direct* erected the marquee. The team were back on site at 7am the next day to complete the antenna work and install the five GB50 stations. At the same time,

117

the RSGB and Burnham Beeches Radio Society set-up 'The Amateur Radio Experience'.

All the transmitting equipment was again kindly loaned by Icom UK. Each station had a modern laptop for logging and was linked to a central server via an Ethernet link. The GB50 station would also demonstrate Automatic Packet Reporting System (APRS), a digital communications protocol for exchanging information among a large number of stations covering a large local area. To improve the experience for the visitor, a large screen showed live details of contacts, with summary data of contacts and countries contacted. Even for the uninitiated, the display created an impression of a modern hobby which had come a long way from the image of Tony Hancock!

The Royal Visit: The highlight of GB50 was the visit by His Royal Highness, the Duke of Edinburgh (Patron of the RSGB), on 3 June. He walked unescorted along the North Terrace and was greeted by Bob G3PJT, RSGB President. Prince Philip was introduced to Paul G3SXE (GB50 project manager) and myself, as well as to Peter Kirby, RSGB

Tim G4VXE explaining the VHF set-up to the Duke of Edinburgh, with Wilf G0WLF and Chris G0FDZ

General Manager. Prince Philip was escorted around the GB50 station by Tim G4VXE (station manager). His Royal Highness spent time at each of the five stations, talking to members of the team and listening to an HF contact with Geoff 9H1EL in Rabat, Malta, where Prince Philip and Princess Elizabeth had spent their honeymoon. Prince Philip also toured the 'The Amateur Radio Experience', meeting RSGB Board members and staff, and students from Harrogate Ladies' College who had contacted the International Space Station. Prince Philip left GB50 later than planned and it was reported on TV that the Royal couple were running late due to an earlier appointment!

118

The Patron's Golden Jubilee message was transmitted to various Commonwealth countries by GB50. The message read:

"As Patron of the Radio Society of Great Britain, I am delighted that it has been able to set up the GB50 Special Event Station on the North Terrace of Windsor Castle overlooking the Thames and the town of Windsor. It is in a very appropriate position to receive messages of good wishes from amateur radio enthusiasts to the Queen in her Jubilee year. I know that the Queen very much appreciates this special contact with people throughout the Commonwealth, and the rest of the world, and she has asked me to send you all her warm thanks for your support and affection at this time. I hope that all your contacts with GB50 over the next 10 days will be 5 and 9. 73 Philip".

In response, messages were received from a number of amateurs in Commonwealth countries, including Ted 5Z4NU, Chairman of the Amateur Radio Society of Kenya, who said:

"We in Kenya always remember that 50 years ago The Queen acceded to the Throne while visiting the Treetops safari lodge near Nyeri, and on the occasion of the Golden Jubilee, The Amateur Radio Society of Kenya offer their congratulations to all concerned in GB50 and wish you all an enjoyable and very successful operation."

Pile-ups and achievements: Before and after public opening hours, the team were able to focus on pile-up operating. At these times GB50 contact rates were well in excess of 200 per hour. Indeed, the location proved exceptionally good for HF propagation. The North Terrace was high above the town of Windsor and provided excellent take-offs to both the USA and Japan.

Propagation at 28MHz was patchy, but the other bands played well with the team amazed to contact 6,790 stations on 15m. But the real success was 6m, where over 50 DXCC entities were contacted. Best DX was PY5CC in Brazil, but other DX included contacts with Morocco, Madeira Is, Ceuta and Melilla, Jan Mayen Is, Jordan, Israel and Cyprus.

The final contact took place at 8pm on 9 June. 52 operators made 24,727 contacts in 145 DXCC entities. All the team's goals were met, both in terms of the number of contacts made and in the number of visitors, and all the equipment and software worked faultlessly.

Team members Tony G3ZRJ, Richard G7GLW, Clare RS102891 and Tony G0OPB (now G2NF) with the crowds at GB50

There were two operating awards associated with GB50, administered once again by Clare RS102891. The two awards – 'The GB50 Golden Jubilee Award', available to anyone who contacted/heard GB50 during the Golden Jubilee weekend, and 'The GB50 Points Award', available for multiple contacts – were another success, with several hundred claims received. Finally, no special event station is complete without a dedicated QSL Manager. Owen G4DFI once again did a sterling job, and still receives QSL cards for GB50 today! Accompanying traditional QSL card requests, Owen received a map of Kobe in Japan, a guide to Winchester Cathedral, and photos of Disneyland, Brighton and a caravan park in Northumberland!

Following the event, a PC CDROM was marketed, articles appeared in five UK magazines and a number of talks were given at various radio society meetings and at the RSGB HF Convention.

After the excitement of GB50…

The www.cvrs.org website gained a new webmaster as Adrian M5ADL took over from Andy M0BXT. The Annual DF Hunt saw four teams trying to find the 'fox' (Richard G8ITB). Some were luckier than others, but 14 members attended the post-DF Hunt gathering and swapped DF Hunt stories. The Society joined with the Darenth Valley and North Kent Radio Societies to enter VHF NFD for the first time in a number of years. The 'Mix and Match' section was entered: Cray Valley organised the 6m station; Darenth Valley the 2m station; and North Kent the 4m station. With no Sporadic-E conditions, G3RCV/P made 90 QSOs on 6m, but it was good enough to obtain 2nd place. G4CW/P was placed 9th on 4m and G0KDV/P 5th on 2m. Overall, the team obtained a very creditable 5th place.

Members were saddened to hear of the passing of Conrad (Steve) G6SDO. A past committee member, Steve had a soft, gentle voice with a wonderful Caribbean twang. He was not too active, but enjoyed operating GB8RE with Dave G4NOW on a number of occasions, and, of course, made all the arrangements for the special event station and visit to Littlebrook 'B' power station several years earlier.

As in 2001, the autumn months were hectic. As a result of the success obtained in the BWBF weekend the previous year, a lower key weekend was arranged for *'Transmission 2002'*. However, using the callsign GB2FB and using the Society's 18AVQ vertical and inverted-V for 40m, the team again raised the most money and won an Alinco DX70TH HF and 6m 100w transceiver. Next up was GB6CW, a special event station for Crown Woods School's Autumn Fair. The following weekend GB6DS, a JOTA station for the 6th Dartford Scouts, was active. A good turnout of Scouts, Cubs, Beavers and Girl Guides made for a successful day. 119 contacts were made in six hours, and there was also an APRS demonstration station and several very popular 'Learn Morse in 15 minutes' sessions.

Lastly in October, a small team – Tim G4DBL, Nobby G0VJG, Ralph 2E0ATY and Richard G7GLW – were active as M8C in the CQWW SSB contest from the 9th Dartford Scout HQ. It was their first major contest together and with strong gale-force winds causing antenna difficulties, were congratulated for a good first effort, making 1,129 contacts. At the same time Paul G3SXE was part of the successful Multi-Multi VC2C Zone 2 CQWW DXpedition team (and again in 2003 as VB2C); Dave G4BUO made 2,819 contacts in the single operator category; and Mick BRS31976, Simon RS177448, Clare RS102891 and I donned our 'London SWL team' hat to log 559 band countries during the same contest. Cray Valley really was active that weekend, in one way or another!

In November, Tim G4DBL aired M8C in CW CQWW. He used an Icom IC-746 into a DJ9SQ vertical loop for 10 – 20m and a 132' inverted 'L' long wire for the low bands, making 600 contacts. The year ended with a traditional Christmas meal, and saw the mid-monthly social re-instated.

2002 was a good year for the Society with increased membership, the success of GB50, several special event stations and some HF and VHF contest operations.

2003

The early 2000s certainly showed a return to successful and sustained activity. Although there was no special-special event in 2003, the year was an extremely busy one for the Society. The year began with another good response from members in the RSGB's AFS SSB contest. Nine members were active, with the 'A' team a little short of the 2002 mark, but achieving 5th place.

The Society's pictorial record of the GB50 event at Windsor Castle was available in March. Working closely with *FDS Graphics* a PC CDROM of the event was produced. It was the first of its kind in amateur radio circles and ran for 30 minutes covering all the major aspects of the event.

A large contingent of members attended the West London Radio and Electronics show to accept an Alinco DX70TH transceiver from Mike G3SED, managing director of *NevadaRadio,* the prize for raising most money – for the second year running – in the BWBF's *Transmission 2002* weekend. This was the first time members wore the Society's new corporate blue polo shirts. A table was hired to promote the Society and sell the GB50 PC CDROM, and new and used equipment on behalf of the BWBF. This raised over £1,000 for the charity.

Dennis G3MXJ (now F5VHY) gave the Society a KW1000 linear amplifier in March as he prepared to move to France. The amplifier was first used by an 11-man team who were active as M8C in the CQWPX SSB contest later that month. Although not as successful as in 2002, due to poor band conditions, 1,515 contacts were made, gaining 4th place in England.

G8JNZ SK

It was with much sadness that the passing of Bob G8JNZ was reported in April. Bob had suffered with cancer for two years. Despite a lengthy stay in hospital, he lost his battle with the disease. He joined the Society in 1963 as a short wave listener, had served on committee and been involved with the production of QUA for 20 years until his illness prevented him from continuing. Bob and I spoke often. He was always totally committed and a real professional when it came to producing QUA. I recall the many times his editorials sought to drive members, unsuccessfully in the main, from their apathy in the 90s. One enduring memory was that on answering his telephone calls, he would always say *"Ahh, there you are"*.

Five teams went out looking for the 'fox' (Richard G8ITB and Owen G4DFI) in the Annual DF Hunt. Hiding in a car park in Joyden's Wood, the 'fox' was found by new winners, a five man team led by Nobby G0VJG, who received 'The Tally Ho! Cup' at the traditional post-DF Hunt social. Nobby also led a group of Society members at the Epsom Radio Fair to raise funds for the BWBF. With the main BWBF fundraiser on holiday, Nobby became the BWBF for the day! Over £500 profit was made, while a separate Society stand made a further £100 for the charity and sold further copies of the GB50 PC CDROM. This event saw the unveiling of a new banner for publicity use at external Society events.

In June, a 'GB50 reunion' took place at our Windsor 'home', *The Carpenter's Arms*. Unfortunately, no date was a good date because many of the GB50 team were either involved in other activities or could not get to Windsor, but nine of the GB50 team met up over a few beers to reminisce on a memorable event for the Society.

The year's first special event station, GB4KME, helped the Eltham-based Kent Model Engineering Society celebrate their 70th anniversary. Six members manned the station, which was organised by Phil G8OPA. A modest station was set up at the Lionel Road Community Centre in Eltham, airing the Society's new Alinco DX70TH and 18AVQ vertical antenna on HF and an FT-817 and ATX vertical on 2m belonging to Simon M3CVN (ex-RS177448). 90 contacts were made in six hours of operation.

After the success of combining resources in 2002, the Society once again joined forces with the Darenth Valley and North Kent societies for VHF NFD, becoming 'The Three Societies' Contest Group' for the weekend. Dave M0BGR and Simon M3CVN organised our part of the event. Operation was from a site at Crockenhill. Cray Valley organised the 6m and 70cm stations, Darenth Valley the 2m station, and North Kent the 4m station. 111 QSOs on 6m netted G3RCV/P 3rd place, but a first serious attempt at 70cm was disappointing. Overall, the combined group achieved 8th place in the Restricted section. Society members Chris G0FDZ, Dave G4BUO and Phil G4EGU were part of the Windmill Contest Group which won the Open section.

In August, the Society won the Restricted section of SSB Field Day. Operating from a site 650' ASL at Cudham, G3RCV/P, running 100w from an FT-1000MP into a dipole antenna, made 725 contacts. Operators were Nobby G0VJG, Richard G7GLW, Ralph 2E0ATY and Simon M3CVN. This was the Society's third SSB Field Day success, winning the *Northumbria Trophy* previously in 1983 and 1984. The team collected the trophy at the RSGB HF and IOTA Convention at the Gatwick Worth Hotel, Crawley, in October 2004.

Operation on behalf of the British Wireless for the Blind

Transmission 2003 marked the BWBF's 75th anniversary, and the Society was invited to operate on their behalf. As the BWBF is a national charity, the special callsign GB75BF was agreed by the RA. Fiona Fountain, head of fund-raising at BWBF said:

"We are thrilled to bits that Cray Valley Radio Society is supporting the work of the British Wireless for the Blind so enthusiastically. BWBF is a totally independent charity established in 1928 and this is our 75th Anniversary Year. Since day one, our mission has been a simple one: to keep blind people in touch with the world, through the wonderful medium of radio…Fund-raising events like Transmission and the support of groups like Cray Valley Radio Society are so important to us."

Operation was from the 9th Dartford Scouts HQ. Icom UK sponsored the event, loaning an Icom IC-756PROII. A second station comprising Ralph 2E0ATY's FT-1000MP and the Society's KW1000 linear amplifier was also used, and GB75BF also had a presence on 2m thanks to Nobby G0VJG's FT-290D. Yagis for HF and VHF and dipoles for 40/80m enabled 921 contacts to be made in 24 hours. A further £920 was raised for the BWBF (effectively £1 per QSO!), taking sponsorship money raised by the Society in three years on behalf of the BWBF to over £4,000.

Richard G8ITB organised the 2003 Christmas social in *'the very posh'* Wickham Suite at the Warren Sports Club in Hayes. With private dining area, lounge and a bar facility, 26 members and guests enjoyed excellent food, and drinks at subsidised prices.

Prior to moving to Devon, Keith G3TAA telephoned to invite me to his New Eltham home to take away any 'junk' which took my fancy. I have to thank him now for all the Cray Valley (and North Kent) newsletters dating back to the 1960s which I took away. Little did I know then how invaluable they would become in writing this history of the Society!

Unusually, the Society's final events of the year were the RSGB 2m AFS and a special event station. Dave G3JJZ organised the Society's entry into AFS, and Richard G7GLW organised GB4RES for 36 scouts from the 1st Royal Eltham group so they could make an amateur radio contact as one element towards qualifying for their 'Communicator Badge'.

2004

The activity theme continued in 2004, but the committee identified a need to cater for all members not just those who enjoyed contests and special activities. It was agreed that at least three meetings during the year should feature a technical talk/discussion. Bernard G8LDV obtained the consent of his close friend's widow and Ofcom (which was formed on 29 December 2003 to inherit the duties that had previously been the responsibility of five different regulators, including the RA) to inherit the callsign G3NPS. Bernard was re-elected as Chairman for a fourth year. As the Society had been running without a Vice Chairman for several years, I filled the post at the Annual General Meeting as Bernard's work commitments meant his attendance at meetings could not be guaranteed. Richard G7GLW was elected as Secretary, and the members of committee were Dave G3JJZ, Nobby G0VJG, Wilf G0WLF and Ralph 2E0ATY. A second 'Activity' booklet (there was one in 2001), edited by Clare RS102891, was issued. It reprised the successful activities undertaken by the Society in 2003. A members' handbook was also sent to all members.

The year saw the Society take part in the newly devised RSGB 80m Club Championship. These were a series of short 90 minute contests with two sections – either 10w or 100w maximum output power – on CW, SSB and Data. They ran between January and July and are still running today. The contests were quite popular amongst members, so much so that when the March SSB leg fell on a meeting evening, the committee actually arranged for six members to enter the contest from home rather than attending the Society meeting! Permission was also obtained from St Mary's Community Complex to erect an antenna at Progress Hall for the evening so Simon M3CVN, with class A supervision, could also air G3RCV in the contest. As many as 11 members per session could be heard on SSB evenings, four on CW evenings, and even two on Data evenings. After seven months of member contest activity, the Society took 5th place in the Club Championship. There was also good member support for the 80m SSB AFS contest. 13 members took part. The 'A' team improved its position again, finishing 4th, compared to 6th in 2002 and 5th in 2003. Simon M3CVN was the top 10w station. Also early in the year, Dave G3JJZ, Simon M3CVN and I (having been licensed since early 2003 as M3RCV) represented the Society in the RSGB's 432MHz AFS contest, being placed 5th.

IOTA 'dry run'

As part of the comprehensive planning for the Society's DXpedition to the Isles of Scilly for the Islands on the Air contest (IOTA) in July, the team of Simon M3CVN, Nobby G0VJG, Ralph 2E0ATY, Richard G7GLW, Chris G0FDZ and Dave G4BUO decided that a 'dry run' was essential. Richard G3YJW loaned his 'annex' and fields in the Kent countryside for the weekend. The team chose an entry into the Multi-two category of the WPX SSB contest as it would provide plenty of band activity, an opportunity to test the equipment under contest conditions, and the ability to erect antennas closely mirroring the expected layout on the island. As the main object was to test everything for the upcoming DXpedition, the team took care over setting up and only hit the bands 15 hours into the contest. However, in just over 24 hours, 1,323 contacts were made, which was enough to secure the team 1st place in England. But more importantly all the equipment and antennas had been thoroughly tested and were ready for the DXpedition. There was also a second, slightly different, 'dry run'. As Nobby was to take all the equipment to the island in his car, the team needed to be sure everything could be squeezed in – the 'dry run' proved that by taking out the back seats, everything fitted – just!

Dave G3JJZ and Pat G0BRV again won the annual DF Hunt and 15 members and guests attended the post-DF Hunt social. Dave also arranged VHF NFD, which saw the Society join forces with the Addiscombe Radio Club to form the 'Kentish Hills Contest Group'. Operation was from the Cudham site used to win SSB Field Day in 2003, but the weather was appalling and so were band conditions. Not surprisingly, the joint team were at the wrong end of the results table – in 9th place in the 'Mix and Match' section.

DXpedition to the Isles of Scilly EU-011

The highlight of 2004 was the first of a number of successful DXpeditions to the Isles of Scilly for the July RSGB IOTA contest. The team – G0VJG, G4BUO, G0FDZ, G7GLW, M3CVN, 2E0ATY and Denise BRS34902 – decided to activate the islands for the contest because they had their own IOTA reference – EU-011 – so those taking part in the contest would need a contact with M8C to gain a new island 'multiplier'. Unfortunately, the team endured a rough crossing on the *Scillonian,* before landing at St. Mary's, locating the site, erecting the 12 antennas, giving the

shack (a farm building) a thorough clean, and setting up the stations. Pre-contest activity was on the WARC bands only, but most of the team 'chilled', enjoying a trip to Tresco, some fishing, the glorious weather and the local ales – but not necessarily in that order!

L-R: Nobby G0VJG, Richard G7GLW and Simon M3CVN operating on the Isles of Scilly

Thanks to the pre-planning, the 24-hour IOTA contest went well. 'Europe 11' was a popular island multiplier, and over 2,300 contacts were made. The team were placed 4th overall in the RSGB 'IOTA contest expeditions' category, scoring 5.6M points, but were awarded the UK High Score Plaque (Expedition). Members were able to obtain a 250 photo CD of the trip. Other members active in the IOTA contest were Tim GJ4DBL/P, Dave G3JJZ, Dave M0BGR, Tony G3ZRJ and myself, winner of the IOTA QRP Plaque.

A busy September

September was a particularly busy month with SSB Field Day and two special event stations – GB2BF and GB2SJS. The same team as in 2003, operating from the same site in Cudham, could not repeat the previous year's SSB Field Day success, but came a creditable 2nd; the contest was lost on 20m, with the 'Contest Cumbria' team – G3IZD/P – making 160 more contacts.

The GB2BF and GB2SJS special event stations fell on the same weekend. GB2BF, was again organised by Nobby G0VJG for the BWBF's *Transmission* weekend. Operating from the 9th Dartford Scouts HQ, the team of Nobby, Ralph 2E0ATY, Simon M3CVN, Richard G7GLW and I used 2 x FT-1000 Mk5s and a VL-1000 linear amplifier into a Force 12 C3SS yagi. Dave G4BUO visited the station on the second day, strung up a dipole for 30m and quickly developed an enormous pile-up. Some contest-style operating netted GB2BF an awesome 3,057 contacts in 43 hours of operation and was able to raise another £1,000 for the BWBF. The Society were the first winners of the BWBF's 'Society Winner' trophy, awarded for the most contacts made during the weekend.

At the same time Phil G8OPA and Bernard G3NPS organised GB2SJS from St. John's Church in Sidcup. Activity was on 40m and 2m. Although a lower key event compared to GB2GF, it was another successful Cray Valley special event.

Tim G4DBL operated M8C for CQWW SSB and CQWW CW. 1,140 contacts were made in the SSB leg and 405 in the CW leg. Dave G4BUO, Ralph 2E0ATY and Simon M3CVN were members of the G1A Multi-One team from Richard G3YJW's QTH, making 3,957 contacts in the 48 hours of the SSB leg. They were placed 2nd in England and 41st in the world.

Richard G7GLW arranged a station on the air evening for a Scouts 'Radio Communicator' badge course. 19 scouts attended and logged stations, learnt Morse and the 'Q' codes, tuned a radio receiver, and passed a 'greetings message'. The final activity of 2004 was the 144MHz AFS contests. Six members took part, being placed 7th. A successful year was concluded with a Christmas meal attended by over 30 members and guests at a new venue, *Ye Olde Leather Bottle* in Erith.

2005

Success came early in 2005 when the Cray Valley 'A' team of Dave G4BUO, Simon M3CVN (operating G3RCV from the 9th Dartford Scout HQ under supervision) and Tim G4DBL won the SSB leg of the RSGB AFS contest. The Contest Committee write-up said *"Congratulations to the Cray Valley RS for achieving what was long thought impossible – taking first place away from the Lichfield ARS."* Indeed, Dave G4BUO was the joint leading station with 352 contacts in the four hours of the contest. The Cray Valley 'B' team were 10th (Nobby G0VJG, G3VLX – with me operating – and Ralph 2E0ATY), and Cray Valley 'C' were 29th (Smudge G3GJW, Wilf G0WLF and Owen G4DFI). This was an excellent start to what was to be another successful year. The Society was also active in CW AFS; the 'A' team was 23rd.

G3XMD SK

The Society's success quickly turned to sadness when the passing of Derek G3XMD was announced in February. Derek joined the Society in 1966. He was soon involved in activities and was elected to the committee in 1967/68. He was Treasurer from 1968/69 to 1971/72 and Chairman from 1972/74. He was one of the Society's most pro-active Chairmen. His enthusiasm rubbed off on others and when the Society was big into contesting in the 1970s, he was the brains behind erecting the antennas. He was also a good contest operator. I can remember logging for him in numerous contests. He particularly enjoyed the stateside runs, often boasting rates of over 100 per hour. However, he was probably best known to members for his auctioneering skills; a 'job' he held for 30 years. Owen G4DFI, Richard G8ITB and I represented the Society at his funeral.

Interest in the 80m Club Championship dropped away after a satisfying first year. The Society was placed 7th in the final listing, some 9,000 points behind the winning society, De Montfort University ARS, who went on to dominate these contests for a number of years.

21 members attended the Annual General Meeting, a much larger percentage of the membership than were attending ordinary meetings at that time! A new Members' Handbook was available at the meeting, listing members and their interests, as well as including a potted history and rules of the Society. Following a proposal by Dave G4NOW, seconded by Nobby G0VJG, at the previous year's Annual General Meeting, an inventory and valuation of Society equipment was available to members for the first time. Members were also advised that due to greater use of the Society callsigns, 2,000 new G1RCV/G3RCV/M8C QSL cards had been purchased (and the supply has not been exhausted 11 years later!). Bernard G3NPS was re-elected Chairman/Treasurer, with Dave G3JJZ filling the Vice Chairman seat. Richard G7GLW remained Secretary. Tony G6LUD was newly elected as a committee member.

The committee gave their support for the planning of two special event stations to be aired later in the year. Wilf G0WLF asked for support to organise a special event station for The Knights Hospitallers of the Sovereign Order of Saint John of Jerusalem, Knights of Malta (an Ecumenical Order). The Order is one of the world's oldest orders of chivalry and was more active in 2005 than at any time in its continuous long history. Even today, the Order's worldwide charitable services carry out an unbroken tradition of its founding charter – to care for the sick and the poor, regardless of race, religion or language. Secondly, Dave G4BUO and I were planning our third special-special event station – a week-long event to celebrate the 200th Anniversary of the Battle of Trafalgar.

With the committee charged to offer a programme which was both activity and technically based, a good mix of meetings were arranged for the year – including lectures by Steve M5BXB on the Vagarda antenna range, a 'project evening' (to replace the defunct 'Construction Contest' evening), and an FT-817 forum. These were interspersed with activity lectures about the second Isles of Scilly DXpedition and Paul G3SXE's contest activity from Canada in 2004.

Ralph was now M0MYC and was immediately involved in planning a G3RCV entry into the ARI DX contest from his home QTH. The team was Ralph, Simon M3CVN, Richard G7GLW and I. 813 contacts led to a 1st place certificate for top score in England.

Richard G8ITB and Owen G4DFI were the 'fox' in the annual, but not too serious, DF Hunt. As it had been accepted that the Society's meeting place was in an RF 'hole', entrants were, for the first time, able to find their own location – no more than a mile from the Society's meeting place – to take a first bearing. It was interesting to note how many entrants chose the same location – high ground adjacent to Eltham Park South.

GB8OSJ was the callsign used for the Knights of Malta special event station at the 9th Dartford Scout HQ. The station was active on HF, 40/80m and 2m. After several years of entering VHF NFD with other local groups, 2005 saw the Society go it alone from the site at Cudham. The team were placed eighth in the Restricted section. The DXpedition to the Isles of Scilly was even more successful than in 2004, with the team winning their section. Nobby G0VJG again organised our BWBF participation and we did even better than in 2004 by winning the 'Society Winner' and 'Most money raised' trophies.

One weekend, two contest successes!

The G3RCV/P team went one better than the previous year in the Restricted section of SSB Field Day. Making over 900 contacts, the team regained 1st place, last occupied in 2003. On the same weekend, a second Cray Valley Group – Paul G4DCV, Tim G4DBL, Bernd M0COH and I won the multi-operator section of the 144MHz Trophy contest operating from Tim's Hampshire hills QTH. The contest was blessed with superb tropospheric conditions for almost the entire contest. Running 300w into a 17 element yagi, the team made 730 contacts for a huge score of 345k. Best DX was SP4SAS at 1,451km. The team were awarded the 'Mitchell-Milling' trophy.

Tim G4DBL operating M8C during the 2005 144MHz trophy contest

Trafalgar celebrations

With the approval of the National Maritime Museum in Greenwich, the Society operated its third special-special event station in five years between 17 and 24 October 2005 as part of the 200th anniversary of the Battle of Trafalgar celebrations.

Once again I was team leader, with Dave G4BUO HF Manager and Chris G0FDZ VHF Manager. We were supported by a truly dedicated team. To mark the occasion, Ofcom issued the Society a third unique callsign – GB200T, and Icom UK again sponsored the station, supplying an extensive range of operating equipment – an IC-7800 and IC-756PROIII for HF and an IC-7400 and IC-910H for VHF. Despite applying for grants, none were forthcoming as they had been for the M2000A and GB50 special event stations. Funding for the event therefore came largely from team members.

GB200T objectives: GB200T was active on all bands from 432 to 3.5MHz daily. The event's objectives were to:

- provide unique publicity for the Trafalgar celebrations through worldwide radio communication

- establish a world-class amateur radio station
- make contact with at least 10,000 stations around the world
- enable members of the public to send 'Trafalgar 200 greetings messages', and
- celebrate the Royal Navy's long association with radio communications.

Station design: The operating station was set-up in a 14' by 10' area inside the E-Library of the Museum. All Cray Valley's special event stations have been set up to enable visitors see the operators face-on when they approach the station – but there are so many examples of special event stations where the public's first uninspiring impression is just the back of the operator's head hunched over a radio! All three GB200T stations were fitted with an audio amplifier and loudspeaker independent of the headset audio so visitors could hear what was going on.

There were three operating positions, HF1 (10 – 20m); HF2/VHF1 (WARC bands and 6m/2m/70cm); and HF3 (40/80m). Due to space limitations, HF and VHF could not be used simultaneously. Dunestar and ICE bandpass filters were used on all the transceivers to minimise inter-station interference. Icom UK also provided a flat screen LCD monitor which was 'slaved' from the display at the centre of the transceiver's front panel. This was used to set the display of the S-meter on both receivers and the band scope, and was positioned so it was in full view of the public.

GB200T used a securely guyed four-section mobile tower to support eight different antennas. All were coax-fed, using almost 1,500' of coax in total. The antenna used on HF1 was a TH5 tri-bander at 65'. Wire dipoles and loops were used on the other HF bands. In true nautical fashion, the tower was *'dressed overall'* with the 40 international signal flags. Meticulous planning before the event and careful labelling of all coax and antennas in advance enabled the whole station configuration, inside and out, to be put together in just three hours.

The tower at GB200T 'dressed overall'

GB200T used 'Starlog' logging software. This product had been successfully used by the Five Star DX Association on the earlier D68C (Comoros) and 9M0C (Spratly Is) DXpeditions. All the contacts at GB200T were fed to a central server, from which daily extracts were e-mailed to the station's webmaster, so radio amateurs around the world could check if they had made it into the log. Society members loaned laptop computers for logging at each station. These were connected on a wireless network. 'Starlog' also enabled the operators to check instantly when someone called to see if they needed a contact on another band or mode. This facility encouraged callers to QSY for a contact on a different band or mode, and also led to an increased number of award claims received by Clare RS102891, who was once again awards manager for the event. The software also had 'talk' facilities so the operators could exchange messages with other operating positions.

GB200T operation: All team members wore corporate purple polo shirts. The station made contact with 13,701 stations during the eight day operation. Apart from many British, European, American and Japanese stations, GB200T also contacted many more exotic locations, including: Alaska, Hawaii, Swaziland, Singapore, Mongolia, Papua New Guinea, Brunei, Hong Kong, Iraq and New Caledonia. A number of special contacts with other 'Trafalgar' stations were arranged for Trafalgar Day on 21 October 2005, including ZL6QH in New Zealand, GB200RN at HM Dockyard, Portsmouth, GB200HNT at HMS Cambria, GB4BOT at Pendeen, Cornwall, and ZB2TRA in Gibraltar.

2,500 people visited GB200T, including those with a family connection to the Battle of Trafalgar. Among them were Anna Tribe, Great, great, great grand-daughter of Lord Nelson; Mary Horatio Arthur, Great, great, great, great grand-daughter of Lord Nelson; and Santiago

Simon M3CVN, Richard G3YJW, and a young visitor passing a 'greetings message', at GB200T

133

Churraca, Great, great, great grandson of Cosme Damian de Churraca y Elorza. Other visitors included Colin White, Chair of the official Nelson Commemoration Committee, Roy Clare, Director of the National Maritime Museum, and Clive Efford, MP for Eltham. 200 younger visitors also passed 'Trafalgar 200 greetings messages'.

The M2000A operation led to an influx of new or re-joining members, including Colin G3SPJ, Cliff G4HSU, Richard G7GLW, Wilf G0WLF and Derek M0BGX.

Swansong: Public enthusiasm for the event was tremendous, the organisation superb and the pile-ups massive, but the lasting memory of the third high profile station organised by the Society was that, once again, amateur radio in the UK had successfully been put firmly in the spotlight by Cray Valley Radio Society. It also provided a unique opportunity for radio amateurs around the world to share in the Trafalgar bi-centenary celebrations.

The three major events – M2000A, GB50 and GB200T – were driven by Dave G4BUO, Chris G0FDZ and me with fantastic support by other team members and the Society. The stations made a total of over 84,000 contacts, and the Society was acknowledged as perhaps the only one in the UK capable of organising and planning such prestigious, high profile, special-special event stations. However, at the end of the eight day event Dave, Chris and I 'retired', leaving the UK special-special event station field to others. But was 'retirement' to be permanent...?

Anyone wishing to know more about GB200T can visit the dedicated website, which is still live at www.gb200t.com

...and finally

Hot on the heels of GB200T was an excellent Multi-two entry in the CQWW SSB contest. A nine-man team took M8C to 1st place in England, 11th in Europe and 30th in the World, with a checked score of 3.27M. Another excellent year for the Society ended with a successful return visit to *Ye Olde Leather Bottle* in Erith for a Christmas meal and social evening.

2006

First Society Foundation licence training course

There had been talk a few years earlier about the Society becoming involved with training for the Foundation licence. Thanks to Chris G0FDZ and Dave G4BUO this came to fruition in 2006 when the Society ran its first training course. Members supported this initiative, with some registering as trainers and others helping out with practical assessments. Eight candidates attended the two-Saturday course, and all eight passed. A great start to the Society's new training initiative.

Practical demonstration by Colin G3SPJ at the Society's first Foundation training course

M8C was aired in the RSGB March 144/432MHz contest. Paul G4DCV, Tim G4DBL, Bernd M0COH and Simon 2E0OAF operated from Tim's QTH on the Hampshire hills. A grim weather forecast had deterred some members from making the trip from the Cray Valley area to Tim's QTH, so a lack of operators led to a 144MHz multi-operator entry only. Given the poor band conditions and freezing temperatures, the 'provincial' Cray Valley team made 341 contacts and achieved 2nd place. M8C was also active on 40m in the ARRL SSB contest from Tim's QTH during the same weekend. 282 contacts achieved a leading G certificate and 9th place in Europe. A team aired G3RCV/P in the UBA contest, being placed 1st in England, and 5th in both Europe and the World.

M8C was also used in the Russian DX Contest (RDXC). Operating from Ralph's QTH, 1,300 contacts were made by the team – Nobby G0VJG, Rob M0CRY, Simon M3CVN, Ralph (now M0MYC) and me. Later in March, M8C was active again in the CQWPX SSB contest. Simon M3CVN drove members on to make a serious attempt at posting a good score. Getting a team together proved quite a challenge, with a number of contesters unavailable at different times during the weekend. However, Simon arranged a rota for a 48-hour Multi-operator, single transmitter entry. The results were pleasing. 2,241 contacts and a claimed score of 4.1M points, led to an all-

2006 CQ Worldwide certificate

time UK high score which stood until 2008. The entry was also placed 14th in Europe and 39th in the World – one of the best major contest results achieved by the Society.

The Reigate Cup was awarded in 2006 as there were sufficient entries to hold a Construction Contest rather than hold a 'Project Evening'. Dave G3JJZ and Dave G3RGS adjudicated the entries and awarded the cup to Colin G3SPJ for his home-brew dual band 2m/6m linear amplifier.

Another non-serious 80m Club Championship season saw the Society finish in 10th place, a long way behind the De Montfort University ARS. However, the Society consistently managed to rally large numbers of members to operate during each of the monthly 2m Club Championship sessions. Indeed, the results showed 37 members active at various times during the year. Competition with the De Montfort ARS remained close until the final session, but a 5th win in 12 sessions sealed a well-earned overall victory, and the G6NB Trophy. Tim G4DBL's consistent performance through the year had much to do with the success.

Annual General Meeting report

A number of important decisions were taken at the 2006 Annual General Meeting. The Society adopted the RSGB's Child Protection guidelines; agreed that the annual accounts should be audited by two elected auditors – Chris G0FDZ and Owen G4DFI were elected for 2006 (and have done the job ever since); announced that the GB200T event had made a small profit and that the 'special event account' would retain a nominal sum to be used for any future special events. It was agreed to publish the Society's asset list. A welcome increase in meeting attendances was noted. Bernard G3NPS was re-elected Chairman, but relinquished the Treasurer's role after 10 years: Colin G3SPJ took over. With membership growing, Dawn M3WTZ was elected as Membership Secretary. Committee posts were filled by Wilf G0WLF, Nobby G0VJG, Ralph M0MYC and me. Adrian M5ADL, who had re-vamped the www.cvrs.org website continued as Webmaster. I took over the editing of QUA (because Simon's exams and Clare's work commitments meant they had little time to devote to the role); and the 2006 Members' Handbook was issued.

Firepower activities

A number of members visited the GB2RA special event station during the Easter holidays at the Firepower Museum, the Museum of the Royal Regiment of Artillery located at the former Royal Arsenal site in Woolwich. The special event was the fore-runner to a more high profile event at the Museum to commemorate the inauguration of the Victoria Cross decoration for supreme valour by Queen Victoria 150 years earlier. The Vintage Operator's Group, led by Cray Valley member, Lawrie G4FAA and Mike M1CCF, asked the Society for its expertise in planning and setting up the station, and to provide operators for the event. A Cray Valley presence led to a more professional event, at which 10,000 contacts were made.

Icom UK again supported the Cray Valley led event by providing IC-756PROIII and IC-910H transceivers. Kenwood also loaned an HF transceiver. Dave G4BUO and Chris G0FDZ advised on, and supervised the erection of the rooftop antennas. The HF stations were located on the floor of the Museum; the 2m station was housed in a *Saracen* army vehicle!

The special-special event callsign GB150VC was obtained. The core team wore burgundy polo shirts, and a number of members operated the station, especially during lunch hours.

The station was in demand, as few GB150 prefixes had been issued by Ofcom. The station was on the air daily between 9am and 5pm. Much pressure was put on the Museum's staff to agree longer, and overnight, hours of operation. Apart from one evening, where operation was agreed from a Land Rover vehicle parked outside the Museum, the requests fell on deaf ears.

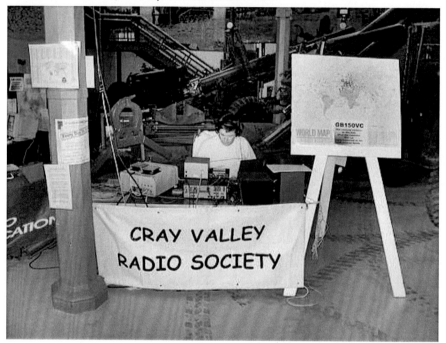

Dave G4BUO operating at GB150VC

Most activity was on 20m SSB and CW, but as the station was operational during the Sporadic-E season, the 6m station enjoyed some good European contacts. Indeed, 531 6m contacts were made in 36 DXCC entities and 147 QTH locator squares. Over the four days of the *'Victoria Cross'* weekend, nearly 3,000 contacts were made, with GB150VC 'spotted' on the DX Cluster over 100 times! As was the norm with Cray Valley special event stations, a successful awards programme was arranged.

More successful activity
The Society entered National Field Day as G3RCV/P and made 1,194 contacts, but was beaten into a close 2nd place in the Restricted section by the Flying Pigs Contest Group (G0IVZ/P) who made seven more contacts. Nevertheless the
138

Society was awarded the Gravesend Trophy for its second place. A third DXpedition to the Isles of Scilly took place for the IOTA contest. The team won the IOTA DXpedition category making 2,965 contacts for a massive 9.5M points total and received two certificates – 1st place Island DXpedition, High Power category, and 2nd place Island, Multi-Operator, High Power Section. In September G3RCV/P entered the Restricted section of SSB Field Day and for the second time in a year achieved second place, but unfortunately the RSGB does not have a trophy available to mark this achievement.

Tim G4DBL arranged another M8C entry for the 2m Trophy contest. Conditions were less favourable than the previous year. 339 contacts earned 6th place in the Multi-operator section. Nobby G0VJG arranged another special event station for the BWBFs *Transmission* 2006 weekend, again making most contacts and receiving the 'Most QSOs' trophy.

Diamond anniversary

The Society was 60 in 2006. Members enjoyed an active and successful month, which saw two excellent lectures - Smudge G3GJW talked about Cray Valley in the '50s and Colin G3SPJ brought things right up-to-date with a talk about WSJT. The third monthly meeting was an Anniversary party which attracted over 80 past and present members and guests. Chris G0FDZ was MC for the evening, and a vast area of memorabilia was on display. Members were also able to relive some past

Owen G4DFI, Lawrie G4FAA and Ray G0FDU at the Diamond anniversary party

glories courtesy of a specially produced DVD. Bernard G3NPS and I addressed the gathering as Chairman and President, before Brian G8OSN, who represented the RSGB, spoke about the Society's past successes. There was sufficient time for 'rag-chewing' and to enjoy the buffet and refreshments, before a toast and cake-cutting ceremony; the cake having been supplied by *J Ayres*, bakers of Sidcup.

The Anniversary month ended with a serious effort in the Multi-two category of the CQWW SSB contest. 4,706 contacts were made for a checked score of 3.98M which brought another 1st place in England certificate, but the score was also 14th in Europe and 24th in the World. Together with the earlier CQWPX result, these two contests were arguably Cray Valley's best results in major international competition.

Throughout the month there was great on-air activity amongst members and non-members as they sought sufficient points to qualify for the Diamond Anniversary plaque. Over 40 plaques were issued. A limited edition 90 minute CD/DVD was available of the month. It included over 300 photos taken by various members during the month-long events. A 60th Anniversary QSL card was also printed for the GB6OCV activity. Anniversary month was followed by a successful Foundation licence course.

The December QUA was sent as usual by post, but also by e-mail – a pre-cursor to the monthly newsletter being sent electronically from January 2007. As a taster, the e-mailed December QUA included colour photographs, to give members an early indication of what they could expect from future issues of the newsletter.

The 'noughties' provided a throwback to the heady 70s, with activity the Society's main theme. Was 2006 as successful as 2000, 2002 or 2005? Opinions varied, but in its own way, 2006 was rather special with the Diamond anniversary and other successful events and activities. However, perhaps the biggest plus was the notable rise in membership and increased attendances at meetings.

2007

Sending QUA by e-mail identified several teething problems, but positive comments outweighed those issues. The decision did not meet with universal satisfaction from members as some still asked to receive QUA by post. The committee honoured that wish, but the eventual aim was to reduce QUA postage costs almost entirely.

2007 saw the first of what is now a regular first meeting of the year – a planning meeting to provide members with an opportunity to influence the committee's direction in terms of the meeting programme and external activities for the year.

The first of the 2007 2m Club Championships in January brought out 16 members: there seemed a resolve to win the G6NB trophy back-to-back! That aim was achieved, with the Society dominating the event and winning 9 of the 12 sessions.

The Society achieved 12th place in the RSGB CW AFS contest (out of 78 entries). The five-man team was G4BUO, G4DBL, M3CVN, G4DCV and G3RGS. Dave G4BUO came 6th in the individual standings, with Simon M3CVN the leading M3. The Society fared better in SSB AFS. The 'A' team of Dave G4BUO, Simon 2E0CVN and Tim G4DBL won the contest for the second time in three years. All three made over 300 contacts during the four hour contest. Dave G4BUO was 2nd in the individual listings. Once again, the Society was the only one to enter four full teams.

The Society's first Intermediate training course was held over three Saturdays in February. It was run by Colin G3SPJ, Dave G4BUO and Chris G0FDZ (with a number of willing volunteers also involved over the three Saturdays of the course). The Society's theme for February was most definitely 'training' as Colin, Dave and Chris explained all aspects of the course at the first meeting, and Brian G8OSN (Guest of Honour at the 60th Anniversary Party) gave a presentation about 'Training and the RSGB' at the mid-monthly meeting. As for the course itself, seven of the eight candidates passed, including Simon (2E0CVN), Kevin (2E0NIM), Karen (2E0KAZ), Alan (2E0LAG), Sam (2E0SDW) and Dawn (2E0WTZ).

March saw a successful construction contest evening with many excellent entries, a mix of exhibits from experienced constructors and construction beginners. The judges had a dilemma in making due allowance for the varying Skill levels of the contestants, without compromising the value of the Reigate Cup. After much deliberation, Sam 2E0SDW was awarded the cup for his VOX switching interface unit, built as an Intermediate course project for use between his PC and rig for auto-CQ calling. The judges were particularly impressed because of the high standard achieved by an inexperienced constructor.

G3VLX SK

Members were saddened to hear of the passing of Deryck G3VLX in April 2007. He was 82. Deryck had been an active member of the Society since 1965. He was Chairman in 1968/69 and 1971/72 and Secretary, Treasurer and committee member for a further six years between 1967 and 1974. From a personal point of view, my memories of Deryck are substantial and vivid. He had much to do with

141

my early years in the Society as an SWL. I remember my first ever half pint of bitter with him and Derek G3XMD after an evening out on Society business; the CQWPX SSB contest DXpedition to France in 1973 as F0RV; the many Sunday afternoons logging as G3VLX/M on the Worked All Britain nets (including one hairy visit to TQ91 Kent across fields and with a few hundred sheep); and closer to his passing, operating for the Society as G3VLX, under his supervision, in AFS and in an 80m Club Championship contest. The Society was represented at his funeral by Dave G4BUO, Richard G8ITB and me. From ideas by Clare RS102891 and Simon 2E0CVN, the committee agreed that the Yaesu FT-900 transceiver and Yaesu FL2100Z linear amplifier donated to the club by Deryck's family would each bear 'In memory' plaques as a lasting memory.

A special event station – GB4VLX – was set up in May from the Dartford Scout HQ in Deryck's memory. Two HF stations used Deryck's equipment as far as was possible, and the event was supplemented by a static display of G3VLX memorabilia, including photographs, articles, his DXCC certificate, and QSL cards from his F0RV, 9H3AM, 4S7VLG, GJ3VLX and GD3VLX

Dave G3RGS operating at GB4VLX

operations. 28 members supported the event. Deryck was also RSGB QSL sub-manager for the G4D and G4R callsigns, and as I had posted out outstanding QSL cards to around 100 G4Ds and G4Rs before the event, many were waiting for an opportunity to work the memorial station. The highlight of the event was the visit by Philip, Deryck's eldest son (later to become M3VJX), who passed some 'greetings messages'. GB4VLX made 1,830 QSOs in 87 DXCC countries in 28 hours of operation. Deryck's memory was celebrated in impressive style. I think he would have been so pleased.

Annual General Meeting
Bernard G3NPS was again re-elected as Chairman. His 2006 review praised the successful training team; my editorship of QUA; the assistance and expertise given

142

to the Vintage Operators' Group which made GB150VC such a success; the gradual peaking of membership; and the contribution newer members were making to the Society's activities. The 2007/08 committee was elected as follows: Secretary – Dawn 2E0WTZ; Treasurer – Colin G3SPJ; Committee members – Alan 2E0LAG, Kevin 2E0NIM, Richard G8ITB and Guy G0UKN. Co-opted members were Cliff G4HSU (Webmaster); Dave G0KPZ (Examinations Secretary); Wilf G0WLF (Trophies Manager); and Simon 2E0CVN (Contests Coordinator). Chris G0FDZ and Owen G4DFI were elected as Auditors for 2007.

It was announced that a sub-committee, comprising Colin G3SPJ, Owen G4DFI and Karen 2E0KAZ were to consider 'tweaking' the Society's rules, following a recommendation from the outgoing committee. Adrian M5ADL relinquished his role as Webmaster; Dave G4BUO and Simon 2E0CVN offered to re-build the www.cvrs.org site. While the website was being re-built, the Society used free web space offered by the RSGB.

Participation in the 80m Club Championships was again disappointing, with the Society finishing 12th. Four of the six 'hounds' found the DF Hunt 'fox' (Richard G8ITB and Owen G4DFI) hiding at Wilmington. 23 members and guests enjoyed the post-DF Hunt social.

Different fortunes in NFD and VHF NFD

Before VHF NFD, Nobby G0VJG secured a field adjoining *The Ship* at Hextable for use in NFD. In perfect weather, Dave G4BUO, Simon 2E0CVN and Dave G3RGS

made 1,140 contacts, which took 2nd place for the second year running. Kevin 2E0NIM enticed non-CW types to attend an excellent late afternoon BBQ.

Simon 2E0CVN and Dave G4BUO operating in NFD

The site for VHF NFD had to be changed at the last minute as the proposed site at Cudham became unavailable. A switch to the NFD site at Hextable, although 140m ASL lower, was arranged. Operation was in the 'Mix and Match' section. Thanks to a donation by Richard G7GLW, the 6m station used a new Trident 5-element yagi, while an 11-element Vagarda yagi at 60' was used on 2m. With concern that the generator would not cope with high power on both 2 and 6m, a decision was made to run the 6m station in the Low Power section using 100w. Guy G0UKN's 'luxury' caravan gave the 2m operators '5 star' accommodation for the weekend, while everyone else made do with camp beds, pump-up mattresses and sleeping bags!

Sporadic-E conditions were present on 6m from the start of the contest. On occasions, using lower power was not too much of a disadvantage, as 124 contacts were made, including some CW stateside contacts and WP4UX (Puerto Rico at 6,813km), who came back to my CQ call! Whereas operating from a lower ASL site had not unduly hindered the 6m station, the 2m and 70cm stations were not so lucky, experiencing poor conditions. The team had a theory that the level of portable activity was affected by the previous week's floods, and because the Tour de France, the Wimbledon tennis finals, and the British Formula 1 Grand Prix were all being shown live on TV! The enforced change of site resulted in a quite poor result, being placed 7th.

A Cray Valley team was again active from the Isles of Scilly for the IOTA contest in July. 3,052 contacts and a 10.1M score in the Multi-operator section won the team a well-deserved trophy.

Eventful SSB Field Day and 2m IARU contest

G3RCV/P was used in SSB Field Day was from a farm location in Knockholt, with the IARU 2m contest entered from the same site. A mobile 60 foot tower supported 2x11 element 2m yagis, with the HF antenna on a 30ft scaffold mast. Everything went well in both events until midnight. It was then that the owner of the farm and his partner drove across the field to visit the team. Having been drinking, and consuming more while on site with the team, the farm owner drove back to his house. 'Crunch'! He drove his partner's Range Rover Discovery 4x4 into a guy stake and kept on going, pulling the tower guys two to three feet and jerking the tower! Fortunately, the other guys took the strain to prevent the tower falling.

144

At dawn, the full extent of the damage to the tower was seen. The tower was still standing, but could not be winched down. Both contests continued, even though the 2m yagis were beaming towards the ground! Dave G4BUO devised a plan to lower the tower, but two other antenna experts, Chris G0FDZ and Dave G8ZZK (together with Phil G4EGU), who were taking part in the IARU 2m contest with the Windmill Contest Group, were summoned for their advice. To ensure there was sufficient daylight to lower the tower safely, the contest operation ended prematurely. The tower was lowered safely, using only a 16ft gin pole!!

Abandoning the contests three hours early cost the team a higher finish in both contests, but 3rd place in SSB Field Day was seen as a good result. In the 2m contest, 276 contacts brought an 8th place finish, using the Gravesend Radio Society's M6Q callsign. This was used because Tim G4DBL, Colin G3SPJ and I used M8C at Tim's Hampshire QTH for the same 2m contest. Unfortunately, Tim's more westerly location was not blessed with the same good tropospheric conditions experienced by stations operating near the east coast of the UK. As a result, M8C were placed 7th.

No quiet end to the year

The Society enjoyed a busy final few months of the year. In October, 22 members and guests made a return visit to the British Vintage Wireless Museum. Sam 2E0SDW organised the 2007 JOTA station for the 3rd Royal Eltham Scouts at their New Eltham premises. The callsign GB50RE was used to mark 50 years of JOTA on HF and VHF. A number of Cub Scouts passed 'greetings messages', learnt their names in phonetics, and completed activities to obtain a 'Communicator' badge. At the same time, Nobby G0VJG organised GB50DS for the 9th Dartford Scouts. Tim G4DBL used M8C from his QTH in the CQWW SSB contest at the end of October as the Dartford Scout HQ was unavailable. 1,525 contacts were made in the Multi-Single category, for 2nd place in England and 43rd in Europe.

A third successful Foundation licence course was completed in November. The team of tutors, helpers and mentors were rewarded with eight candidates applying for an M3 callsign. The Society hired three tables at the Kempton Park rally in November to sell members' surplus equipment, which saw a £200 profit for Society funds.

At this time, the Society used Nasko LZ1YE to design and print its QSL cards. He was certainly kept busy. In 2007 alone, he designed and printed a new four cards

– G3RCV/G1RCV/M8C, G3RCV/P and M8C (for the Isles of Scilly DXpedition), GB150VC and GB4VLX: 5,500 cards in total.

Recalling the generator issue at VHF NFD, the committee resolved purchase of a sturdier machine for use at field events, with members invited to donate £50 towards the cost. The committee was clear, however, that if members contributed financially, they would not 'partially own' the generator: it would be owned by the Society.   It is also interesting to note from committee minutes late in the year that issuing QUA electronically had saved the Society £341 in running costs during the year, and that there had been discussions about the possibility of organising a special-special event station for the 2012 London Olympic and Paralympic Games.

Finally in 2007, over 30 members and guests attended the Christmas meal at *Ye Olde Leather Bottle* in Erith. Attendance at the pre-Christmas drinks evening at the *Bull* on Shooters Hill was decimated by illness, however those who attended believed the venue was the best to be chosen for the pre-Christmas drinks social for some years.

2008

Although Cray Valley led the field after the first 2m Club Championship of the year and never relinquished it, the committee accepted Simon 2E0CVN's recommendation that the Society should take no serious part in the 80m Club Championship contests because the format of combined SSB, CW and Data contests did not play into the strengths of the Society. This decision did not prevent members from taking part if they wished to. There was also no serious attempt at the year's 80m AFS SSB contest

The Advanced training team: Frank G0FDP (Examinations secretary), Dave G3RGS, Dave G4BUO, Chris G0FDZ, Colin G3SPJ and Richard G8ITB with successful candidates Sam M0SJW, Karen M0KAZ and Kevin M0KSJ

because it clashed with one of the days of the Society's first Advanced training course for the Full licence. For the record, Tim M0AFJ, Owen G4DFI, Smudge G3GJW and I came 24th of 89 entrants. The Advanced course training team were Chris G0FDZ, Dave G4BUO, Colin G3SPJ, Richard G8ITB and Dave G3RGS. Thanks were also extended to the examination invigilation team of Dave G0KPZ (Examinations Secretary), Tom G1FAD and Joyce, Colin's XYL. All three candidates were successful - Karen (M0KAZ), Kevin (M0KSJ) and Sam (M0SJW). Following their success, Karen and Sam donated the 'Sandell-Whitehead' trophy to the Society to be awarded to the candidate at future courses deemed to have demonstrated exceptional effort and enthusiasm.

The Society entered the CQ WPX SSB contest in March but things did not work out as planned. Nobby G0VJG was holidaying in Antigua as V25V, so manpower issues caused a late start. There was also a 'Murphy's Law' incident with a borrowed linear amplifier, and the team had to finish the contest eight hours early because there were insufficient members to help take down the antennas after the contest. However, 1,042 contacts were made in 21 hours for a claimed score of 1.045m. Bob VE3SRE, Peter M3PHP and Paul M3JFM were part of the team for the weekend.

A business meeting month

April was a business meeting month. Notice was given of an Extra-ordinary General Meeting to discuss a revised set of Society rules, and this was followed later in the month by the Annual General Meeting. The proposed new rules built on the excellent revised rules that Nigel G1BUO oversaw as Chairman in the 90s. The exercise was largely to update and pave the way for a healthy and active Society for the next 10 years. The meeting was a quiet affair.

Between these meetings, Colin G3SPJ and I collected the G6NB Trophy at the 'Radio Fairs' rally at Kempton Park racecourse for the Society's success in winning the 2007 2m Club Championship.

As the Society moved towards the Annual General Meeting, the committee noted a much firmer financial footing, thanks to Colin's G3SPJ's stewardship. A surplus had been accrued, which was spent during the year on improving training, special event and contesting facilities. I chaired the AGM as Bernard G3NPS was absent because of a recent knee operation. The meeting was well-attended and also passed off quietly. There was little change to the committee, with Keith G4JED replacing Alan 2E0LAG, who declined a further year on committee for work reasons.

A new special event venture

The annual 'Mills on the Air' weekend had been running for some years bringing together radio clubs and the Society for the Protection of Ancient Buildings. As a new venture for the Society, a special event station was established at Meopham windmill in Kent. The callsign GB6MW was chosen as the windmill was of a six-sided smock design. The event was organised by Kevin M0KSJ. There was something for everyone – radio operating; windmill heritage; chatting to members, relatives and the public; press photo calls; good food, beer and glorious sunshine!

In total, 404 contacts were made (330 on the HF bands and 74 on VHF), including 22 of the 39 mills taking part in the weekend event. The most exotic contact was a classic piece of good timing (HS0ZEE in Thailand on CW) as Dave G4BUO came across his

Tony G6LUD, Kevin M0KSJ and Karen G8JNZ at Meopham Windmill

signal while being interviewed by the press! Of course, VHF activity was much more local.

The Windmill Trust was pleasantly overwhelmed by the scale of the unfolding event and were impressed by the professionalism shown by the Society and its members. GB6MW certainly resulted in the busiest ever National Mills weekend for the Trust.

New top-of-the-range generator purchased

After considerable research and discussion, the committee decided to purchase a top quality product from the respected market leader. The generator chosen was the Honda EU26i. Its appeal was the very quiet, fully enclosed design, which used a multi-pole alternator to produce DC to then power an internal inverter. The advantage of this arrangement was that the inverter produced a 230V 50Hz pure sine wave output with an exceptionally good voltage regulation of +/- 1%. The 2.4kW continuous power rating was sufficient to run one complete 400 watt station easily, or three separate 100 watt stations required for a full, Restricted section, VHF NFD entry. The new generator was first used during the NFD weekend at which the Society also entered the UKSMG 6 metre contest. As expected, the generator powered both stations perfectly and ran for almost 27 hours on a little over 20 litres of fuel. The purchase, with the help of some member donations, was the most expensive item owned by the Society at that time, but the committee believed it was a sound purchase and one that would prove its worth over many years.

A field activity summer

The Society was involved in a triple event in June from the *Bull* site at Hextable. As well as NFD and the UK Six Metre Group contest, Kevin provided a summer BBQ with a menu of spare ribs, chicken tikka kebabs, burgers and sausages and assorted greenery, bread with relishes – all for £5!

Kevin M0KSJ preparing the summer barbeque at the NFD/UKSMG contests weekend

Dave G4BUO organised NFD. He, Mark M0DXR and Simon 2E0CVN used a Yaesu FT-1000MP transceiver to a wire antenna. HF conditions were poor, especially in the final two hours of the contest. G3RCV/P made 1,198 contacts and was placed 4th out of 33 entrants in the Restricted section, but were the 80m band winners. Overall, it was a close run thing as there were only 203 points between the winners, the North of Scotland Contest Group, and G3RCV/P in 4th. Indeed, there were only 65 points covering 2nd, 3rd and 4th places.

As Colin G3SPJ and I were members of the UK Six Metre Group, a decision was taken to enter the group's 6m contest, which took place on the same weekend as NFD (although starting and finishing earlier). Colin's Yaesu FT920 and the Society's 5-element yagi were used. The band had been open for Sporadic-E contacts before the contest, but conditions petered out shortly before the contest. A diet of UK stations followed, until the Sporadic-E returned, when some good contacts into the Balkans were made. An early start the following morning paid dividends with several Turkish stations and the best DX – JY4NE (at over 3,500k). G1RCV/P made 104 QSOs. It was not a winning score, but it was a good test run for VHF NFD.

Additionally, but not a Society event, Kevin M0KSJ and Tony G6LUD set up a 2m QRP station for the 3w Practical Wireless contest which ran through the Sunday. The station consisted of an Icom IC-7000 running into a 5-element lightweight Sotabeam. In just over two and half hours of leisurely operating, 32 contacts were made, ranging from IN88 in the South (Brittany) to IO74 in the North (Kirkcudbrightshire in Scotland) together with stations in Wales and Yorkshire. To work Scotland using 3W and a Sotabeam was a revelation to Kevin and Tony!

Hot on the heels of the NFD weekend was VHF NFD. There were a number of plusses. Kevin M0KSJ arranged for the use of a site at Wrotham which was 200m ASL. Colin G3SPJ, Guy G0UKN and Richard G8ITB did much of the preparatory work and loaned most of the equipment, and Simon 2E0CVN masterminded the antenna work. The only real minus was the inclement weather that arrived a few hours before 'teardown' which meant the team got soaked! From an operating point of view, the entry was in the Restricted section, and the weekend provided the best VHF NFD result for many years. Overall, G3RCV/P was placed 3rd, only 150 points shy of 2nd and 700 short of winning outright. The 6m station, with Mark M0DXR making the majority of contacts, won the Restricted section; the Icom IC-756PROIII transceiver and Trident 5-element yagi performing faultlessly.

Guy G0UKN, Colin G3SPJ and I achieved 3rd place on 2m, making 250 contacts using an Icom IC-7400 transceiver to an 11-element yagi. On 70cm, an Icom IC-7000 transceiver with a high performance SSB Electronics mast-head pre-amp and 21-element yagi was used; a number of members made 77 contacts for another 3rd place. For 4m, Guy G0UKN serviced an old Europa transverter, which used valves for the transmit section and he built the necessary high voltage PSU especially for the event. The transverter was driven at 28 MHz from a Yaesu FT920 transceiver. Unfortunately, after the hard work to provide a 4m presence, the results were disappointing.

A busy autumn

Looking back at the activity reports in QUA, September 2008 was a particularly busy month. However, I begin, with a report of something a little different! The late August BBQ coincided with a Quiz Night at the *Bull* on Shooters Hill and an eight-person 'Cray Valley Brain Cells' team pitted their wits against some other knowledgeable teams. A thoroughly entertaining evening saw the team obtain 3rd place.

Both the SSB Field Day and 2m Trophy contests were well supported from the *Ship* site in Hextable. SSB Field Day saw an entry into the Open section for the first time. Using Dave G4BUO's trusty TET tribander on the tower that was damaged in 2007 and a 40m delta loop and inverted-V for 80m at 55ft, the contest was a success with 1,460 contacts made. G3RCV/P took 1st place by a 20,000 point margin from the East Notts Contest Group to collect the *Northumbria Trophy*. Poor weather and the low ASL *Ship* site led to a poor result (8th out of 9) in the 2m Trophy contest.

Members had again been taking the 2m Club Championship seriously, and the October success – for the fifth month running – retained the G6NB Trophy for the third consecutive year (2006, 2007 and 2008) with two months of the competition still to run. Some members attended the trophy presentation at the Kempton Park rally in April 2009. Members were also active in the RSGB 80m Sprint contests which took place between August and December. Cray Valley was top of the 'also rans', in 5th place 28,000 points behind the winners, De Montfort University ARS, who continued their domination of 80m contests.

There was no serious entry into the CQWW SSB contest in 2008. However, Nobby G0VJG and Andre M3XNL aired M8C from Nobby G0VJG's QTH making 475 contacts. The Society was also represented in the results by Dave G4BUO, who was part of the successful 6Y1V (Jamaica) contest team, while I achieved the 80m QRP record for England (which someone bettered the following year!). Owen G4DFI achieved 1st place in England in the Low Power Single Operator, All Band section, and also 1st place in the Tribander/wires, Low Power SSB All Band section of the same contest. In this latter section he was 4th in Europe and 12th in the world.

Special event stations: Guy G0UKN organised the GB2AF special event station which operated from the British Vintage Wireless Museum in Dulwich. The event celebrated 100 years of radio from the building. From 1908 to 1914 a teacher from Dulwich College, Alfred Rickard Taylor, 2AF, lived there. He was a wireless pioneer and early member of the RSGB. In 1914, the house was bought by the father of Gerry Wells, and Gerry developed a passion for radio which lasted until his death early in 2015 aged 85. One of the stations at the three day event used vintage equipment: KW Vanguard transmitter, Eddystone EA12 receiver, KW201 receiver, Johnson Matchbox ATU and doublet. Power output: 30w carrier, anode and screen modulated AM. There was so much nostalgia for the radio amateurs who were lucky enough to contact the station, some of whom had contacted 2AF in the 1930s. Two other stations, using more modern equipment, made just over 700 contacts using the GB2AF callsign.

Guy G0UKN at the GB2AF vintage radio station

Nobby G0VJG and I were invited to attend the BWBF's 80th anniversary celebrations. The event also marked the launch of their new *Concerto* radio, and their new premises in Maidstone. We were in good company, meeting Ed 'Stewpot' Stewart and 'Whispering' Bob Harris who officially launched the new *Concerto* radio. We were pressed to take part in *Transmission* 2008 to provide some competition for the 'Most QSOs' and 'Most money raised' trophies. Of course, we accepted! So GB2BF was active again from the 9th Dartford Scouts HQ in October. However, with other special events around the same time and few members available to help with setting up, it was a more modest event than had been the case in previous years, but 650 contacts were made in less than a day, which won the Society the 'Most QSOs' trophy again. A further £725 was raised in sponsorship for the BWBF charity.

Guy G0UKN also organised the JOTA station for the 1st Royal Eltham Scouts, GB2RE. Although the event marked 100 years of scouting in Eltham, Ofcom did not approve the application to use GB100RE. The event took place at the District Headquarters in New Eltham. The shack had a constant stream of visitors and many Scouts took the opportunity to pass 'greetings messages' to other Scout groups around the British Isles and the world. Some Scouts studied for their 'Communicator' badge by learning callsigns, licensing and propagation, logged stations and tuned a communications receiver. Just under 200 contacts were made over two days, but the politeness and enthusiasm of the Scouts really impressed the team and gave the weekend a really good buzz.

The end of another year...

But not before the Society attended the Kempton Rally to sell surplus equipment, and the December Foundation training course led to 10 new licensees. 35 members and guests partied at *Ye Olde Leather Bottle* before Christmas. The raffle, which made a £230 profit for Society funds, saw Matthew Roberts (Guy G0UKN's son) win the evening's star prize of a 2m handheld on the day before he was successful in the Foundation Licence assessment. To conclude an amazing few days, he was also awarded the 'Sandell-Whitehead' trophy following the course as the candidate who the tutors believed had demonstrated the most effort and enthusiasm. Matthew was licensed as M6MER (and is currently 2E0MER).

2009

After the humdrum of the previous year, 2009 began with entries into the RSGB AFS contests. In the CW contest, the Society unusually entered a 4-man team – Dave G4BUO, Simon 2E0CVN, Mark M0DXR and Paul G4BXT: 14th place was achieved. In the SSB leg, members again showed their affinity with the contest, losing out on top spot by only nine contacts. Only a last minute fault with his amplifier forced Mark M0DXR to run with 100w – without 'Murphy' striking, a Cray Valley first would appear to have been a formality. The 'A' team comprised G3RCV (which I operated from Richard G3YJW's QTH), Dave G4BUO, Nobby G0VJG and Mark M0DXR. The 'B' team of G3GJW (operated by Richard G7GLW), Guy G0UKN, Paul G4BXT and Owen G4DFI were placed 17th. The 'C' team of Chris G0FDZ, Kevin M0KSJ, Nigel G1BUO and Cliff G4HSU were 42nd, but were better placed than 28 'A' teams.

I had more than a passing interest in the Society's Intermediate training course; I was one of the candidates! From a personal point of view, the training team did an exceptional job and all seven candidates passed; I became 2E0RCV.

Cray Valley sweat shirts and fleeces were introduced early in the year to supplement the corporate blue polo shirts introduced several years earlier. Members were able to represent the Society at external events during the less warm months of the year and still maintain a corporate image.

Early Olympic planning

Members were advised early in the year that some groundwork had been done to bring amateur radio to the 2012 London Olympic and Paralympic Games. Dave G4BUO, Mark M0DXR and I had given two well-received Powerpoint presentations to the London Organising Committee of the Olympic Games (LOCOG). The RSGB endorsed the proposals following a further presentation to their Board. This resulted in a seat on their Olympic sub-committee. Dave and I also attended an invitation only meeting at the O2, at which Lord Coe spoke.

With eight Olympic events taking place in Greenwich borough, and with the experience gained in three special-special event stations, the case had been made for the main amateur radio event in England to be organised by the Society.

However, this was the start of a very long, and often bumpy, road, even within the Society itself.

Annual General Meeting

According to QUA, this was *"the lowest profile Annual General Meeting ever"*. The question was asked if providing too much advance information was the reason, or if it was because the committee had done such a great job in running the Society, such that attendance was deemed unnecessary?

The meeting had one notable moment as Bernard G3NPS announced, after he had again been re-elected as Chairman, that he would be taking a well-earned break, after some 30 years, from committee business after the 2010 meeting. A paper ballot was required to decide the Vice-Chairman, as both Kevin M0KSJ and Guy G0UKN volunteered for the post. The vote was close, but Kevin shaded it. The other major change saw Dave G4BUO take over role of Programme Secretary. I had looked after the Society's programme for quite a few years, so the change was well overdue.

The 'Worked Cray Valley' award was re-launched in June with five levels of achievement; anyone claiming the 'Gold' award (for achieving 100 points in contacting Society members, club stations and special event stations organised by the Society) would be able to choose their own trophy (within reason)!

Summer external events and success in NFD

After a great deal of planning by Kevin M0KSJ, GB6MW was operational for a second year at Meopham Windmill for the 'Mills on the Air' weekend. The event was well-supported by members but the weather was not so kind – it was definitely an event where the new Cray Valley sweatshirts and fleeces were an essential part of the kit! The event was given some good publicity by *Radio Kent* and the local press, and the team enjoyed some relaxed operating on 40m SSB and 2m FM and SSB. For the Sunday public Open Day, an HF yagi was erected in the mill grounds to enable a second HF station to be assembled in a tent outside, together with a display area for amateur radio handouts and training material. This was popular with the public and visiting amateurs.

The Society's entry into NFD was to the low power 10w, 12 hour, section, largely because Dave G4BUO had injured his back and could not sit for too long! Simon 2E0CVN and Mark M0DXR did most of the operating, making 605 contacts. This brought a 1st place finish, beating the previous low power score by some margin

Mark M0DXR operating in NFD 2009

(and it is still a record today); the team won The Reading Trophy. Members not involved in the contest joined the contesters to enjoy a BBQ, and some contacts were also made in the UK Six Metre contest.

Maritime Mobile to the Thames forts

Some members enjoyed a 'cruise' down the River Thames to the Red Sands and Shivering Sands forts in the Thames Estuary, activating G3RCV/MM on the way.

The Red Sands fort

The Thames sea forts are the last in a long history of British marine defences. Designed by British civil engineer Guy Maunsell (1884-1961), the seven tower army anti-aircraft forts played a significant role in World War 2. In the 1960s, the forts were home to pirate broadcast radio stations; Radio Invicta (later King Radio and then Radio 390) broadcast from Red Sands fort, and Radio Sutch (later to become Radio City) broadcast from the Shivering Sands fort. More information about the forts and their preservation can be found at http://www.project-redsand.com/

The G3RCV/MM trip went smoothly enough but venturing into the shipping lanes off the Kent coast proved a real eye opener! There were reports of large sea-going vessels moving at speed in all directions with the 'RCV boat effectively becoming the *'rabbit in their headlights'!* Operation was on HF (81 contacts) and VHF (18 contacts).

While the Cray Valley team were engaged in matters maritime, Guy G0UKN and Matthew M6MER were taking part in 'Roper Family Camp', a weekend break at Downe Scout activity centre near Biggin Hill. This was an annual event for members, former members and hangers on of the 1st Royal Eltham Roper pack cubs. It was not usually a radio event, but Guy and Matthew were keen to contact G3RCV/MM. They set up a Yaesu FT-857 with battery power into an 80m inverted-V dipole. I ran the Society 80m net for the /MM operation. This was the first time three 'RCV' callsigns had been on the air at the same time.

Nigel G1BUO and Paul G3SXE proved to be a very elusive 'fox' in the annual DF Hunt. Parked in a *'bit of a hole'* in the car park of the Bexley Sports Ground at Baldwyn's Park, they were only found late in the day by two teams; four other teams were unable to DF the 'fox'. The Reigate Cup was presented to Frank G3WMR and Dave G4YIB at the post-DF Hunt social.

...and VHF NFD success, too

VHF NFD in 2009 was an unqualified success. Operating as G3RCV/P on four bands from a site situated 215m ASL at Fairseat, near Sevenoaks, the Society bettered its 2008 3rd place by coming 1st in the Restricted section, to be awarded the Martlesham Trophy. Much of the success was attributed to being in the right place at the right time to milk the superb Sporadic-E conditions which the contest, unusually, enjoyed. Mark M0DXR did most of the 6m operating and basked in Es propagation for almost the whole contest. He also caught a multi-hop Es opening to the Caribbean in the evening which brought over 14,000 points from just two contacts into 9Y4 (Trinidad and Tobago). 165 contacts and a score of over 148,000 points won the 6m Restricted section by some margin.

Although I operated the 6m station for the final two hours of the contest, I majored on 2m and found two separate Es openings; the nine stations contacted in Spain, Portugal and the Canary Islands were worth 16,000 points. The contact with EB8BRZ at 2,907km was the best DX worked by any station in the Restricted section. G3RCV/P was 2nd in the section.

The overall success was achieved by steady performances on 70cm and 4m to supplement the fine performances on 2 and 6m. The Society's newly acquired 7-element 4m yagi led to an improved score on that band, and the DVK (Digital Voice Keyer) had much to do with the 96 contacts on 70cm.

Keith G4JED at the 70cm VHF NFD station

The relatively small team (Andre 2E0BUU, Colin G3SPJ, Guy G0UKN, Keith G4JED, Kevin M0KSJ, Mark M0DXR, Nobby G0VJG, Richard G8ITB and me) worked hard to erect the masts, assemble the stations, make 565 contacts on all bands over the 24 hours, and clear the site after the contest. A hard weekend, but a very successful one!

A footnote to the 2009 VHF NFD was the visit of 'Rustus' (aka Vigo Royal Rustus), a pedigree Ayrshire bull calf, to within 15 feet of the 6m tent. After a lengthy stand-off with three of the team, he retreated to join the rest of the herd, but it had not been an easy 'meeting'!

Between 30 July and 6 August 2009 in excess of 3,500 Scouts, Guides, Explorer Scouts, Cubs and Brownies from all over the world gathered at the Guides Association site in Cudham for the Kent International Jamboree. Members, led by Richard G8ITB, Keith G4JED and Sam M0SDW, helped set up a special event station, GB4KJ. The purpose of the well-equipped station was to educate the participants about amateur radio and enable the passing of authorised 'greetings messages' to other Scout and Guide stations worldwide, as well as taking group lecture sessions about amateur radio as a hobby and its links within Scouting and Guiding.

Summer socials

Two summer socials were arranged, both at the *Bull* on Shooters Hill. The weather intervened when 25 members and guests had assembled for an outdoor

158

BBQ. Not to be outdone, the food was cooked and served on the premises and everyone enjoyed a pleasant evening.

Some of the GB4SH team: L-R: Chris G0FDZ, Tom G1FAD, Cliff G4HSU, Frank G4WNF, Wilf G0WLF and Colin G3SPJ

Over the August Bank Holiday weekend, Cliff G4HSU arranged a special event station, GB4SH, to celebrate 'Festival Summit' weekend. Dating back to 1749, the pub was reputed to be haunted as well as being in contention for the highest pub (above sea level) in London. Each year, the landlord offered local groups his garden space to promote their hobby or activity. The Society set up HF and VHF stations and made a casual 150 contacts, but promotional opportunities were, sadly, limited.

Sad losses

The sad deaths of Paul G4BXT and ex-member Frank G3FIJ were reported during the summer of 2009. Paul first joined the Society in the 1970s as an SWL and operated several contests with the Society, including VHF Field Day on Nine Barrow Down in Dorset. His interest in radio declined, but he re-joined the Society shortly before his death, operating in both the 2009 AFS contests. Frank Howe MBE had been a past member of the Society, but founded the Colchester Amateur Radio Club. He took part in various contests for the Society and regularly submitted scores for the annual 'Worked' tables. In recognition of his services to the community, Frank was presented with an MBE by the Queen at Buckingham Palace in 1998.

SSB Field Day and 2m Club Championship success

The first weekend in September saw entries into SSB Field Day and the 2m Trophy contest from the *Ship* site at Hextable. The SSB Field Day operating team of G7GLW, G0UKN, G0VJG, G4TSH, G4BUO and me retained the *Northumbria Trophy*, again beating the East Notts Contest Group. Although less contacts were made (1,118 compared to 1,313), the team found 14 more multipliers and that proved decisive. This success meant that the Society had won all three field days contests it had entered in 2009 - SSB Field Day (Open), National Field Day (QRP) and VHF National Field Day (Restricted). Was that a first?

While the HF field day contest was in full swing, others aired M8C in the 2m Trophy contest using Nobby G0VJG's new 18-element M2 yagi. Knowing the site was not an ideal VHF contest location beforehand, the M8C airing ensured that the Society catered for members who wanted to operate in, arguably, the premier VHF contest of the year. Using only 100w in the Open section, the result was as expected: 130 contacts for 7th place.

While the year had seen some exceptional external activity, members had been doing their best to retain – for a fourth year – the 2m Club Championship. After a poor start, the Society clawed back the Trowbridge & District ARC's lead and moved ahead with a few months of competition left. With wins in each of the June to December contests, the Society comfortably won the G6NB Trophy.

CQ Worldwide contests: M8C appreciation

As another decade in the Society's history is concluded, it is worth noting that M8C had been aired consistently in the CQWW contests during the 2000s. A trawl through the on-line scores database provided this dedicated breakdown of M8C involvement, with particular successes noted in 2005 and 2006.

Year	Mode	Category	Checked score	Contacts (QSOs)	Position - England	Position - Europe	Position - World
2002	CW	SO LP All	218,286	544	11	229	432
2002	SSB	Multi-One High	611,665	1,082	7	93	199
2003	CW	SO HP All	796,656	1,498	9	80	220
2003	SSB	Multi-One High	523,110	1,248	8	84	169
2004	CW	SO HP All	122,008	405	12	158	427
2004	SSB	SO HP All	711,632	1,140	3	53	204
2005	CW	SO HP 20m	336.774	1,509	2	16	32
2005	SSB	Multi-Two	3,274,475	3,881	1	11	30
2006	CW	SA HP All	57,936	372	8	148	416
2006	SSB	Multi-Two	3,989,580	4,706	1	14	24
2007	SSB	SO HP All	865,920	1,525	2	43	111
2008	SSB	Multi-One High	137,946	451	11	127	248
2009	SSB	Multi-One High	164,206	571	8	134	263

Non-activity matters

Behind the scenes, members were livid about proposals for the centralised marking of Foundation and Intermediate exam papers, and the temporary withdrawal of Instructor registrations for several members due to 'a major administrative error' concerning CRB checks. The committee pursued both issues hard.

The committee also considered a request to start meetings at 7.30pm rather than the traditional time of 8pm because of the longer distances some members travelled. After due consideration, it was agreed there were more advantages in maintaining an 8pm start, but it was agreed that meetings should start promptly at 8pm, that the tea break should remain at 9pm but for only 15 minutes, and that the raffle would be drawn at 9.55pm. This arrangement proved satisfactory and still applies today.

All 10 candidates from the December Foundation licence course were successful. By popular choice, the Society's 10m Sunday morning net was moved forward one hour, to 10.00am, towards the end of the year and continues to be well-supported today.

Chapter 8: On the home straight...the recent past:
2010 to the 70th anniversary

I have reached the home straight in my nostalgic journey through *'The Cray Valley Years'*. The last few years have been eventful. Although the Society's finest (and last) major special event station for the 2012 London Olympic and Paralympic Games was an unparalleled success, it caused friction amongst certain members. The Society found a new home, but internal disputes within the committee also led to some resignations and uncertain episodes. The new home has since led to the establishment of a shack with a fully-operational Society station with HF and VHF yagi antennas and dipoles for members' use. And it is fitting that the conclusion of this 70 year journey ended with a magnificent anniversary party which attracted over 100 members, past members and invited guests.

2010

11 members were active in the RSGB AFS SSB contest, but the 'A' team could only manage 4th place due to a series of mishaps. Five members took part in the RSGB 70cm AFS contest. Nobby G0VJG, Colin G3SPJ and I made up the three-man team 'A' team which came 7th. George G3BNE and Guy G0UKN were also active in the contest.

The Society ran its second Advanced exam course for the Full licence over three weekends in January. There were six candidates, including 2E0RCV! And everyone passed! New callsigns were later announced as: Jasper M0JDA, Bob M0MCV, Andre M0PIA, Jim M0OOD, Ian M0UAT and Billy M0ZWW.

The Girl Guides celebrated their 100th anniversary in 2010, and to mark the occasion Ofcom issued the very special prefix GG100 for Guide stations to use during the 'Thinking Day on the Air' weekend. The Society was approached by the

Royal Eltham Girl Guides to run a station; the callsign GG100RE was obtained to use at the Eltham United Reformed Church, Court Road, Eltham (where the Society used to meet more than 30 years ago). 12 Girl Guide stations were contacted out of a total of 284 contacts. A number of Guide leaders and Brownies passed 'greetings messages', while Smudge G3GJW, with a couple of Morse keys and oscillators, taught the Brownies and Guides the art of CW.

Nobby G0VJG with Brownies passing a greetings message at GG150RE with Frank G0FDP and Andre 2E0BUU looking on

There were 13 entries to the year's Annual Construction Contest, showing that construction was still alive in Cray Valley. Kevin M0KSJ and Ian G0AFH judged the exhibits. Certificates were awarded to Matthew M6MER for a PSK31 interface, Ian M0UAT for a light modulated Tx/Rx system, and to Slim G4IPZ for a 70cm HB9CV antenna. The Reigate Cup was awarded to Malcolm G8MCA for a bespoke designed and built gel battery charge unit.

Bernard G3NPS finally retires

With notice duly given of his planned retirement at the 2009 Annual General Meeting, Bernard finally 'retired' after 10 years as Society Chairman and 35 years on committee. Following his retirement at the Annual General Meeting, Bernard said:

"As a club we have come a long way in those 35 years and during my time as Chairman and on the committee we have risen to one of the premier amateur radio societies in the country. I would like to take this opportunity of thanking you all for your support over the years and to wish Kevin M0KSJ every success in his new role."

Kevin M0KSJ was elected Chairman. His message to members was:

"Little did I know as a curious 10 year old, tuning around a dusty old jumble sale valve radio and hearing the banter of a Cray Valley top band net, that one day some 40+ years later not only would I be a club member, but I'd end up being elected Chairman! Well here I am... still coming to terms with what I've taken on."

By AGM standards, the meeting was low key. Prior circulation of the 'bumper bundle' provided members with much information, especially about Society finances and an up-to-date and robust statement of current owned equipment. The meeting discussed the suitability of Progress Hall as a meeting venue in light of bigger attendances and the rumoured suggestion that budget cuts by Greenwich Council could impact on our tenure at the hall.

Cray Valley Award re-launched

The May QUA saw the long-awaited re-launch of 'The Cray Valley Award'. The award had been claimed by many since its inception in the 1960s, but had lain dormant for too many years. The 2009/10 committee believed the time was right to re-launch the award. Kevin M0KSJ, with photographs provided by Chris G0FDZ, Cliff G4HSU and Colin G3SPJ, spent time honing it into a colourful parchment in keeping with modern times. To help publicise the award, an activity week was arranged. Six awards were quickly issued.

More 2010 activity

Nobby G0VJG and Ralph M0MYC aired M8C in the CQWPX SSB contest from Nobby's home QTH making 550 contacts. Kevin M0KSJ organised the third GB6MW activity from Meopham Windmill for the 'Mills on the Air' weekend. However, the weather for mid-May was not for the faint hearted as temperatures failed to climb out of single figures, with a blustery wind and rain showers. Operators complained of the cold as the draught whistled through the door taking the feeder cables to the operating position on the first floor of the mill. The Mill Trust took pity and provided several fan heaters before handing over the key to the mill (an original handmade key about six inches long that Quasimodo would have been proud of!), and a fire alarm key with a piece of paper with 'reset instructions'.

Fire alarm! Yes, the mill had a new fire alarm installed and the mill folk were a little concerned about how it operated, having never seen it in action – adding that in the event of a false alarm, we had *"7 seconds"* to reset it before the flashing blue lights set out. It was also noticed that the alarm control panel was radio-linked to the smoke sensors and panic buttons. Stories of alarms and the fire brigade at Rangers House in the early hours of New Years' Day 2000 were recounted, but thankfully the alarm stayed silent and the GB6MW operation went smoothly over the two days. With almost 50 mills on the air, activity was fast and furious with 284 contacts made on HF and another 76 on 2m. Our presence, helped by press, magazine and local radio coverage resulted in many local amateurs visiting the station. As a bonus, Chris G0FDZ organised one-off personal guided tours to the GB3VHF 2m beacon site located close to the mill.

The 2009 VHF NFD success was only followed by a 4th place finish in 2010. This was largely due to the Sporadic-E 'footprint' not extending far enough or for long enough on 6m, and not reaching 2m at all! G3RCV/P was, however, one of only 10 stations in the 6m Restricted section to work over 3,000km into Cyprus. 3rd place was achieved on both 2m and 70cm. Although 9A6R (Croatia) was the best DX on 4m, G3RCV/P only achieved 10th place on that band.

The DF Hunt was blessed with good weather. Paul G3SXE and Nigel G1BUO were once again the 'fox', and four teams of 'hounds' chased them to *The Mary Rose* pub in St Mary's Cray. But being close to railway arches, a number of bearings were later found to be incorrect, such that several teams headed off in the wrong direction! The cryptic clues were also of little help, as signals from the 'fox' were so poor that many of the clues could not be copied! Only two teams found the 'fox', with Guy G0UKN and Matthew M6MER presented with 'The Tally Ho! Cup' during the post-DF Hunt social.

The Isles of Scilly team in 2010 was Dave G4BUO, Chris G0FDZ, Nobby G0VJG, Rich G7GLW, Keith G4JED and Karen (who by this time, with the approval of Bob's parents and Ofcom, had inherited Bob G8JNZ's callsign). The main reason for the trip, as ever, was to activate EU-011 in the RSGB IOTA contest.

Chilling on the beach: Chris G0FDZ, Rich G7GLW, Nobby G0VJG and Keith G4JED

It was a smaller-scale operation than previous years, with 100w transceivers and dipole antennas. Even though band conditions were poor, the team still won a trophy for leading low power British Isles DXpedition. Away from the radio, trips were taken to Tresco Abbey Gardens and Bryher. EU-011 was particularly sought after post-contest – a fact that bemused the team!

After a gap of several years, the Society once again took part in the 48-hour charity event – *Transmission 2010* – as GB2BF to support BWBF's work to supply specially adapted radios and audio equipment to the blind and partially sighted. This was to be the last *Transmission*. Over the years, the Society had raised over £6,000 for the charity, spurred on by Nobby G0VJG's encouragement and hard work. The 2010 event was organised by Kevin M0KSJ. Operation was from a venue new to the Society - the Margaret McMillan House field study centre at Fairseat, Kent. Members supported the event with vigour, making a total of 741 contacts and raising a further £950 for the charity. This qualified the Society for the 'Most QSOs' and 'Most money raised' trophies for the final time.

Andre M0PIA and Nobby G0VJG drove the mobile trailer tower to the *Ship* at Hextable in readiness for the SSB field day weekend. Set up was well-supported and the contest itself went well. G3RCV/P made more contacts and worked more multipliers than previously, for a winning score of 998k points. The success was the Society's third consecutive SSB field day victory, and was even sweeter for beating the Bristol Contest Group, operating as G6YB/P, who had more contacts. But once again, the key to the success was the number of country multipliers worked. Congratulations were extended to the operating team of Nobby G0VJG, Mark M0DXR, Dave G4BUO and Rich G7GLW for retaining the *Northumbria Trophy*. While the operating team shared the glory, they were extremely grateful to those members who helped set up and teardown. Unfortunately, a check log had to be submitted for the 2m trophy contest which took place at the same site, because it was later found that one of the operators was not an RSGB member.

Drawing the year to a close

Having concentrated on the busy summer/autumn activity, I must refer to the unveiling of a new range of Cray Valley clothing. The garments were of a heavier industrial quality, compared to the previous light leisurewear. Displaying the Society logo, the fleece was in dark blue and the polo shirt and sweatshirt in dark Royal Blue. Members were able to personalise the garments by having their name and callsign embroidered. This clothing range is still the corporate face of the Society today.

Chris G0FDZ was the first recipient of the 'Gold' Cray Valley Award for obtaining 75 points in contacting member, Society and special event stations organised by the Society since 1 January 2000. Members had been taking part in the 2m Club Championship during the year, but it was the Bolton Wireless Society and Trowbridge & District ARS who took first and second places. No fifth success, unfortunately. The Bolton club really took the event seriously, achieving 11 consecutive monthly victories from February – December (after Cray Valley won in January). The Society was 3rd, some 1,300 points behind the Trowbridge club, having achieved second place in the final four contests.

2011

2011 began with belated Christmas activities due to the disruption caused by unusually heavy snowfall early in December 2010. Many members had their work and personal plans thrown into chaos because of the 8 – 12 inch snowfall, and Society activities suffered too. The main December meeting was postponed, however a decision was taken to proceed with the Foundation Course, although one candidate and several of the tutorial team were unable to make their way through the snow to the exam venue. Five candidates were pleased the course went ahead, as their new M6 callsigns were announced as – John M6ALW, Phil M6HIQ, Lakai M6LKH, Marlene M6MLZ and John M6RWH.

The Christmas meal, scheduled for early December, was also a weather casualty, postponed until mid-January. Once again, the raffle arranged by Clare RS102891 was the highlight of the evening, with Richard G8ITB arranging the star prize of a 2m FM transceiver, which was won for the second year running by the Kenney (FAD) household! The snow and ice had largely melted enabling the Christmas mid-monthly meeting at the *Bull* to take place. Over 20 members attended and enjoyed a superb selection of beers and wines, and a free buffet. A donated raffle was arranged, thanks to the Chairman, whose very fine Christmas hamper was the star prize. Originally meant as the star prize at the postponed Christmas meal, the hamper had to be raffled as certain perishable items would have been 'past their sell by dates' if it had been held over for the January Christmas meal!

The January Intermediate course was a success with nine of the 10 candidates successful. New callsigns were announced as: John 2E0SOL, Adrian 2E0MUS, Patrick 2E0PBF, Jim 2E0JWG, John 2E0TBK, David 2E0KHZ, Lakai 2E0TBU, Marlene 2E0MLZ and John 2E0DTA. The unsuccessful candidate from the December 2010 Foundation exam also made no mistake second-time around. Once again, the training team of 20 volunteers were rewarded by the successes. The Cray Valley 'model' is based on members delivering exam module presentations and others organising and running practical sessions. Members also mentor the candidates to help them better understand aspects of a syllabus that could be quite demanding for those trying to find a rung on the amateur ladder or progress up it. Some members attend the courses simply to listen and learn. The whole operation being expertly held together by a course lead instructor, and made all the more enjoyable by just the right number of tea, coffee, biscuit and cake breaks!

AFS 2011

There was a keenness to arrange a competitive 'A' team entry to the RSGB AFS SSB contest, as the Society's last success had been in 2007. The challenge was to put arguably the top four contesters in a position where they each had a competitive station to try to post a competitive score. Nobby G0VJG operated from the Dartford Scouts HQ with Ralph M0MYC and Rob M0CRY. Dave G3RGS hosted Mark M0DXR, Dave G4BUO operated from his home QTH, and I operated from Frank G4WNF's QTH. The effort was justified as Cray Valley romped to victory! G0VJG and I were 6th and 9th, with 346 and 328 contacts respectively. M0DXR was 11th and G4BUO 17th. The team won the *Flight Refuelling ARS trophy*. The 'B' team of Malcolm G8MCA, Chris G0FDZ, Owen G4DFI and Wilf G0WLF were placed a creditable 29th. Dave G4BUO, Ian G0AFH and Nobby G0VJG entered logs for the earlier AFS CW contest, and made a total of 345 contacts.

Cray Valley reflector introduced

With greater reliance on e-mail-based Society communications, often with large file attachments, the use of the member's e-mail distribution list showed signs of strain. An optional, alternative, way to receive Society e-mails and QUA was introduced – a member only Internet Yahoo Group (an e-mail reflector). To increase security, the group was not traceable in Yahoo's public searchable index and the contents were hidden from public view with access controlled by the Society. As well as dealing with e-mail traffic, the group site served as an on-line storage area for sharing information to subscribed members. So rather than take up space on a member's hard drive, files like QUA were stored to be accessible to members when needed, but they could also be downloaded to a local hard drive or directly printed off, as required. This was an optional enhancement, but approximately half the Society's members subscribed, although even today some are technically challenged to fully enjoy the facility.

The Annual General Meeting was well-attended and passed off without too much controversy or any difficult questions from the membership. Richard G8ITB had already announced his intention to stand down as Secretary; Malcolm G8MCA took on the role, with Frank G4WNF joining the committee.

The main May meeting was held at the 1st Royal Eltham Scouts hall off Southend Crescent, Eltham – where the Society's training courses were held. The committee arranged the move to the larger hall so the evening could combine two activities – demonstrations of the Society's newly purchased Icom IC756PROIII transceiver and the new Society reflector. This was a pre-cursor to a change of meeting venue.

2011 external activity

There was a disappointingly low turnout to HMS Belfast's Easter Activity Week visit. Marc G0TOC gave up his time to provide a first class guided tour of the ship, moored in the Pool of London, and an opportunity to operate the permanent special event station on the ship, GB2RN. One of the more interesting facts from the evening was that a major restoration project to rebuild replacement masts was unveiled by HRH The Duke of Edinburgh and former Arctic Convoy veterans during the previous October. HMS Belfast is, of course, one of the few surviving Royal Navy ships that served in the Arctic convoys, helping to keep Russia supplied and able to fight Hitler's armies during the Second World War.

The fourth return to Meopham Windmill as GB6MW for the National Mills Weekend appeared to re-kindle members' enthusiasm for special events as a number of newly licensed members took the opportunity to operate the station. Record numbers of mills took part in the event from the UK and further afield in Belgium, Holland and Ireland.

The year's DF Hunt and social were well supported. Paul G3SXE and Nigel G1BUO were again the 'fox'. They were located in Woolwich - off the Common and near to the Royal Artillery Barracks. Two teams were successful in locating them – Guy

169

G0UKN and Matthew 2E0MER, and Frank G3WMR and Dave G4YIB. As for the other 'hounds', it was a mixture of frustration, wrong bearings, and a failure to understand the clues! A subtle rule change – 15 minutes between transmissions from the 'fox' – led to a more relaxed tour around the Greenwich streets.

The Society enjoyed a day at the 'Greenwich Great Get Together' on land adjacent to the Royal Artillery barracks at Woolwich in June. The event coincided with Armed Forces Day. The 'Get Together' was arranged by Greenwich Council and our invitation came through the Greenwich Action for Voluntary Service (GAVS), of which we were, and still are, a member. Although the special event

The GB2GGT QSL card

callsign GB2GGT was obtained, the main activity was to publicise the Society. A very professional pre-dressed Marler-Hayley display was set up, together with a fine QSL card display, leaflets about the Society, its courses and literature publicising the planned 2012 Olympic activity. The 'Get Together' was well-attended, although not too many of the visitors to the Cray Valley stand, including the Mayor of Greenwich, had ever heard of amateur radio!

VHF NFD was again from the Platt House Farm site at Fairseat, but the entry was not competitive, although the 4m station had the best Restricted section DX - GM4ZUK/P at 653km.

New meeting venue announced

It was announced that after 26 years, the Society would be moving its meetings from Progress Hall, Eltham to the 1st Royal Eltham Scouts HQ off Southend Crescent, Eltham from September. With meetings consistently attracting upwards of 25 members, the committee believed the Society had outgrown the Progress Hall facilities. Ties with the St Mary's Community Complex continued, however, with committee meetings continuing to be held at Progress Hall. Many members were already familiar with the Scout Hall as it was one of our registered exam centres used for Foundation and Intermediate licence training courses. It was on this basis, coupled with outgrowing space at Progress Hall, that the

170

committee pursued the idea of a mutually beneficial coming together for a new club meeting venue. The new location offered considerably more space for meetings, had a refurbished kitchen area and an adjacent annex room which had the potential for several activities to run in parallel, even the possibility of airing G3RCV. On a more practical note, potential storage space could be utilised, and the Scout committee was keen to develop a closer working relationship. The new venue was also better served by local public transport, although onsite parking was (and remains) limited.

How times change for the better, as Graeme G6CSY and I negotiated the move to Progress Hall from the Christchurch Centre in Eltham High Street in 1985 because of a reducing membership and falling attendances. The Society met at Progress Hall over 500 times, hosting some fine lectures, but few will remember the escaped mice at a surplus sale, Jim Bacon G3YLA's (yes, 'the BBC weather man') lecture about VHF propagation, Ted G3DCC's 'travelogues' or Dud Charman's G6CJ's 'Antenna Circus'? Great times.

65th anniversary

The Society celebrated its 65th anniversary in October 2011. The committee decided against holding a celebratory evening, but the Society callsigns were aired. GX3RCV was active from the 9th Dartford Scout HQ, followed by a presence at 'Apple Day' at Woodlands Farm, Shooters Hill. Guy G0UKN, Owen G4DFI and I aired G3RCV to help members and non-members work towards the 'Worked Cray Valley' award. Operating G3RCV, I found out at first-hand how highly regarded the Society was. These two comments summed up the general view - *"We've all heard of Cray Valley and read about all your successes in RadCom"* and *"Most people envy Cray Valley Radio Society"*. The comments were not from members or friends of the Society, but from radio amateurs unknown to me and who did not live in the Cray Valley catchment area. I think the comments show how well Cray Valley Radio Society is held in esteem by its fellow radio amateurs.

More activity and a note of sadness

SSB Field Day was referred to in QUA as *'…We all had a good time but made a few mistakes!'* The results appear to bear that out as G3RCV/P was 4th in the Open section over 60,000 points adrift of winners, the Bristol Contest Group, G6YB/P.

The Society had a presence at the West London Radio and Electronics Show at Kempton Park in November. Three tables were manned with items unsold from the earlier surplus sale and other equipment. The day netted a £300 profit. A couple of weeks later saw another early start for the Society's seventh Foundation Course. All 10 candidates achieved a first time pass. The Christmas meal was arranged by Kevin M0KSJ, with Guy G0UKN organising a raffle featuring some excellent donated prizes. 24 members attended the pre-Christmas meeting at the *Bull* on Shooters Hill.

All through the year, members had been taking part in the 2m (and 4m, 6m and 70cm) Club Championship but not with the gusto of previous years due perhaps to rule changes. The Society was 4th, some way behind the Bolton Wireless Club, who won the 2m, 4m, 6m and 70cm sections.

A note of sadness came at the end of the year with news that Patrick 2E0PBF had lost his battle with cancer. Patrick came to amateur radio late in life and obtained his Foundation and Intermediate licences at Society training courses, a point mentioned with some pride by his daughter in his eulogy. Patrick had hoped to complete the hat trick and attain his full licence, but unfortunately with failing health that was not to be.

2012

The year was dominated by 2012L, the special-special event callsign granted by Ofcom to celebrate the London 2012 Olympic and Paralympic Games, but it was far from an easy ride. More on this later, but let's look at the traditional Society calendar first.

The first activity of 2012 saw 13 members submit logs for the SSB leg of the RSGB AFS contest. Unfortunately, the 'A' team's success of the previous year could not be repeated, with the team in 3rd place. The 'B' team was placed 24th place, and the 'C' team 64th from 90 entries.

The 2012 Construction Contest was a well-supported event with a dozen entries. Malcolm G8MCA was awarded the Reigate Cup for his home designed and built PIC digital SWR meter. 'Commended' certificates were awarded to Tony G6LUD, Slim G4IPZ and Des G6WCX.

Guy G0UKN was elected Chairman at the Annual General Meeting. Kevin M0KSJ took over the Treasurer's role from Colin G3SPJ and the Vice-Chairman role was left vacant. The committee had two new members for 2012/13, Des G6WCX and John 2E0TBK.

The Society again supported National Mills on the Air weekend with GB6MW at Meopham Windmill. The HF station was an Icom IC-756 ProIII with a KW1000 linear amplifier installed on the upper level of the mill as in previous years, and a

Yaesu FT857 for 2m and an AnyTone 5189 for 4m on the ground floor. A 'keep it simple' principle was adopted for the antennas - a doublet for the LF bands hung from the mill cap and assorted dipoles for the higher bands. VHF was catered for using a vertical 2m colinear, a 9-element 2m Tonna and a 4m homebrew 'Slim Jim'. The weekend was a success on the operating front – with over 300 contacts made – and also on the public relations side, with a good number of visitors.

Faces at GB6MW 2012

By comparison, the DF Hunt was poorly supported. Nigel G1BUO and Paul G3SXE were again the 'fox', but the 'hounds' endured a frustrating evening. 'Team

Roberts' – Guy G0UKN, Matthew 2E0MER and Paul M6BHU – were the only team to locate the 'fox', their third successful 'chase'.

VHF NFD was again from the Platt Farm site at Fairseat, but it was again a lower key effort, mainly due to a lack of personnel. The team had an enjoyable, but tiring weekend, with G3RCV/P placed 6th in the Restricted section.

G3BNE SK

George G3BNE joined the Society late in life. Born in 1922, he volunteered as aircrew in the Royal Air Force but was rejected because he wore spectacles, but his knowledge of physics and maths enabled him to train as a ground wireless fitter and he was later selected for Radar training. For the rest of the war he worked on coastal height-finding radars and was promoted to flight sergeant at the age of 21. On D-day plus one he finished the war in Antwerp where he met his future xyl, and after a whirlwind courtship of three weeks they were married, for 63 years. Following demobilisation in 1946, George was employed by Decca where he rose to head of department retiring in the 1980's. His interest in amateur radio began after the conflict. His equipment was home brew and initially CW moving to AM later. In the 1960's with the advent of SSB and HF beams he abandoned HF and spent the rest of his radio life exclusively on 2m and 70cm. He was a keen contester and always maintained meticulous records of the countries and locator squares worked and confirmed. These are now part of the Society's archives. George died in August 2012 at the age of 90. His memory was celebrated by a new trophy, awarded to recognise operation in the 2m UKAC contests, of which he was very keen.

More activity and December events

For the second year, the Society manned a special event station for 'Apple Day' at Woodlands Farm, Shooters Hill. The special callsign GB4WFT was aired on HF and VHF, but more emphasis was placed on explaining amateur radio to visitors. Kevin M0KSJ provided an interesting array of publicity information to help promote the educational value of amateur radio.

Instead of taking a stand at the Kempton Park rally, the Society took one at the Coulsdon Amateur Transmitting Society's 'Bazaar' for the first time. Five members manned the stand, complete with a new Society banner.
A high pass rate was achieved in the Society's eighth Foundation Course at the end of the year. Following a suggestion from Smudge G3GJW, the committee

agreed that, starting with the December 2012 course, exam candidates would be offered a year's free Cray Valley membership on completion of their course. The initiative enabled newcomers to get a firmer foothold in the hobby by attending meetings, taking part in events, getting help with choosing a transceiver and antennas, as well as providing the opportunity to put their questions to members as they took their first steps into their new hobby. The free year's membership also extended to any Intermediate candidates who were not already a member of the Society.

The 2012 Christmas meal was held at the White Hart Carvery and Restaurant in Eltham Hill, Eltham early in December. The highlight was the raffle, which saw Karen G8JNZ win the star prize of a 2m handheld. Frank G0FDP won the Chairman's prize, while Owen G4DFI won the President's prize of a magnum of award winning Côte du Rhone wine. Later in the month, there was a good turnout for the traditional Christmas drinks evening. The Christmas morning nets were also well-supported with 15 members on the 80m net and 20 on the 2m net.

2O12L at the London Olympic and Paralympic Games

It was recorded in QUA in 2009 that early discussions had taken place for the Society to organise a prestigious special-special event station to celebrate the London 2012 Summer Olympics. The ride was not easy!

As with previous special-special event stations, a core team was assembled as a sub-committee to oversee the detailed arrangements. The goal was to operate from Rangers House, the location for M2000A. However, through Clive Efford, MP for Eltham, the Chief Executive of English Heritage advised that the venue was unavailable, but gave permission for the team to discuss their plans with the local manager at Eltham Palace. While developing detailed plans with English Heritage, there were further meetings with the London Organising Committee of the Olympic Games (LOCOG) and Ofcom, and agreement was secured for the use of the 2O12 prefix (Two Oscar Twelve). The hope was to have a flagship station in each of the four home countries to spread the Olympic experience around the country and to replicate *'Team GB'*. However, much as the organising team tried, only the Barry Radio Club enthusiastically took up the challenge and successfully organised a 'sister' event, 2O12W for Wales.

When the going gets tough..... Unlike the 2O12W team, who secured funding from the Welsh Assembly and the National Lottery, the Society's carefully prepared applications for grants were unsuccessful because there were so many competing Olympic sports projects in the London area. This immediately led to concern within the Society's main committee about funding, insurance and liability issues, such that one member resigned.

Another major headache for the core team was the stringent branding restrictions imposed by LOCOG. These were put in place to protect sponsor's brands and broadcasting rights, affecting every athlete, Olympics ticket holder and business in the UK. As well as introducing an additional layer of protection around the word 'Olympics', the five-rings symbol and the Games' mottoes, the major change in the legislation was to outlaw unauthorised 'association'. This barred non-sponsors from employing images or wording that might suggest too close a link with the Games. These restrictions led Icom UK to withdraw their offer to supply equipment to the project, causing another headache. However, LOCOG did agree that the project could sign-up to the 'London 2012 Inspire Programme', which had been established to recognise outstanding non-commercial projects genuinely inspired by the London 2012 Games. Agreement meant that 2O12L literature (including QSL cards) could display the 'Inspire' mark, enabling the project to use the promotional tool to connect with the Games and reach out to its audience.

It was not until a meeting in early March 2012 that the main Society committee agreed unanimously that the project could go ahead without limited company status, but instead with the services of a nominated Health and Safety officer. A revised budget, submitted by the sub-committee, was also approved as the committee believed there were sufficient funds, pledges and promises to proceed with the project. At this point Dave G4BUO and I each donated £1,500 to the project, and George G3BNE, sadly to become SK before the event began, donated £500. Other members kindly donated to the project – G3GJW, G3RGS, G3SXE, G4IRN, G4NOW, G8IPY, G0UKN, M0BGR, M0JXG, and the Society itself. Others raised funds by selling donated equipment – G3SPJ, G8ITB, G4HSU, G8MCA, G0FDP, G0FDZ, M0KSJ and 2E0OCD. Another valuable source of income was the sale of 2O12L polo shirts. As the Olympic Games began, income for the 2O12L project stood at over £7,200, slightly more than the projected budget spend total.

But funding proved to be something of a minor headache, as it was at about this time that an impasse had been reached with the local authority over planning permission for the masts to support the antennas. Yet another headache to resolve! The organising team went from one challenge to another, but the

176

bitterest pill came when, after a meeting with English Heritage's Head of London Historic Properties, approval was refused to proceed with the project at Eltham Palace. The team were devastated! All the work done with the local English Heritage manager, who it is believed overstepped his authority, was lost during a 45 minute meeting! The reasons for the decision were largely commercial:

i) *the visual impact of the masts was inappropriate and unsuitable for the story being told at Eltham Palace about 1930s splendour, fashion and opulence*

ii) *the masts and cable runs would seriously compromise the ability to generate additional hospitality and filming revenue*

iii) *the existence of so much valuable electronic equipment would present a serious security risk for the Palace, and*

iv) *using one of the meeting rooms* (to house 2012L) *was highly undesirable for English Heritage.*

Plan B? Resilience had nothing on the 2012L organising team! Through the Society's excellent links with the local Scouts, within two weeks of the English Heritage decision, agreement had been reached to use the Greenwich District Scout Activity Centre in New Eltham for the 47 days of the 2012L operation (between 25 July and 9 September 2012) . Although this affected 'footfall', it had two immediate benefits:

i) *reduced concerns about budget, and*

ii) *less complications about the antenna farm, and substantially more room to erect antennas!*

HF Manager Dave G4BUO was delighted! He was able to plan how the project could utilise the additional space for the HF antennas, and VHF Manager Chris G0FDZ was quick to plan an increase in the size of the VHF yagis. QUA reported that it felt as though *"The starter's gun had fired and we were on our way to the first bend"*. However, planning permission issues with the local authority ran until the 11th hour. Having consulted the RSGB's Planning Advisory Committee for advice, an application had been submitted for a Certificate of Lawfulness. Although the antennas could have remained without any permission for the whole of the Olympic Games, they would have contravened planning law had they remained in the period until and during the Paralympic Games. It was not until the day before the Olympic Games began, and with the antennas erected, that the local authority agreed verbally to endorse the required certificate. Interestingly, the certificate did not arrive until early September!

While the lack of Icom UK's sponsorship initially caused some concern, the Society, its members, other radio amateurs and supporters – Neville G3NUG, Nigel G3TXF, the Chiltern DX Club, and the Newbury and District ARS in particular – donated everything required to be able to run five competitive stations.

The antenna farm at 2O12L

The 2O12L exhibition: A very important component of 2O12L was the exhibition. Kevin M0KSJ arranged this superbly, designing a number of posters and displays himself, and also reviving the set of nine large display boards used at M2000A twelve years earlier. Most of the 2O12L display material was in a conservatory area adjacent to the main operating room, and visitors were able to view a display of basic transceivers and a model house displaying a typical

Part of the extensive exhibition area at 2O12L

domestic amateur radio station with doublet antenna and VHF yagi (see opposite). Beyond the exhibition area, Smudge G3GJW and Dave M0BGR ran 'Learn Morse in 10 minutes' sessions, teaching visitors to send their names in Morse code; successful candidates were awarded a certificate. **Foolproof operation:** Knowing that dozens of different operators would use the station at some time during the 47 days of operation, band changing was made as simple and foolproof as possible. The HF1 and HF2 stations were fully automated, with a change of band on the rig also changing the band in the Win-Test logging program, changing the Dunestar bandpass filter and selecting the correct antenna via the 6-Pak switch located outside. On the WARC band station (HF3) it was necessary to select the filters and tune the amplifier manually. Logging software was Win-Test in DXpedition mode, which performed flawlessly throughout.

Before the enforced move of QTH, much work had been done on operator numbers. In addition to Society members, invitations were issued to known DXpedition and contest operators in the UK to help operate the station. However, with the move to the Scout centre it was realised that while it was a great benefit to be able to operate around the clock, this brought a requirement for even more operators. It was a great surprise that, with a few exceptions, the take-up of invitations experienced in pile-up operating was extremely poor. It is still a puzzle today why someone would spend several thousand pounds travelling half way round the world when they could have experienced comparable pile-ups for a mere £26, to cover the insurance cost and the purchase of a team polo shirt. One operator who approached us early on in the operation was Fred G3SVK. He did sterling service as one of the few CW operators, putting a huge number of contacts into the log, especially on 17m. Fred remains a valuable member of the Society today.

The opening ceremony: All previous prestigious special event stations organised by the Society have featured a launch party. This one was no different. Guests of honour were the Mayor and Mayoress of Greenwich, MP for Eltham Clive Efford, local community and Scouting representatives and the core 2012L team. A buffet was provided, followed by the customary welcome speeches and the cutting of a celebratory 2012L cake by local Member of Parliament, Clive Efford (see opposite). With the formalities complete, a pre-

arranged 'sked' with the Welsh Olympic station, 2012W, had been arranged. However, 40m band conditions did not favour 'inter-G' working. Instead, Don G3BJ, who was well-positioned near to Wenlock Edge where the first Games of the modern Olympiad took place, was able to exchange 'greetings messages' on 80m with Clive Efford, the Mayor of the Royal Borough of Greenwich, and Greenwich Council's Games liaison officer.

2012L hits the bands! After the inaugural contacts, 2012L began transmissions simultaneously on five bands. 350 contacts were in the log after one hour, with the first thousand contacts taking only 2 hours 40 minutes. 3,300 contacts were made by midnight. The pile-ups were immense, making a nonsense of the 'poor' band conditions. Late on day 2, 17m conditions to the west coast of the USA and Japan were phenomenal, and with a 2-element yagi, the pile-ups on 40m were

huge. Day 2 also saw the first 6m Sporadic-E (Es) opening – to Scandinavia. Although near the end of a poor Es season, 2012L made 500 contacts in 43 DXCC entities and 176 locator squares. Best 6m DX was an SSB contact with the D64K Comoros DXpedition. As the operation progressed and the pile-ups on the HF bands subsided, some less experienced operators had a lot of fun using a station they could only dream about having at home. In total, 106 operators made contacts at 2012L.

Mick M0XBF, who made many of the 6m QSOs at 2012L

As with all the major special event stations the Society had organised, 2012L generated a very high level of interest in Japan, with almost 2,000 JA stations logged on 17m alone. Throughout the operation daily contact numbers and DXCC entities worked were tracked. These charts and tables provided a topic of interest to visitors. Also tracked were how many countries competing at the Olympics had been contacted: a Cray Valley version of the *'Olympic Truce'*. As expected, it proved hard to contact many of the competing African nations, but 167 participating countries were contacted.

Each day saw a mix of UK and DX radio amateurs visit 2O12L, including A71AM, NH7C and ZL1TOU. Among the UK visitors were Graham Coomber G0NBI, General Manager of the RSGB, numerous Ofcom staff who visited after a day working at various London Olympic venues, and Eric G8GP, who had reached his 100th birthday earlier in the year. Several radio societies also arranged visits.

RSGB General Manager, Graham G0NBI, operating 2O12L

The start of the Paralympic Games saw a second launch party. This included a presentation by the BWBF and a pre-arranged contact with the RAIBC club station, followed by one-to-one contacts between several disabled Society and RAIBC members.

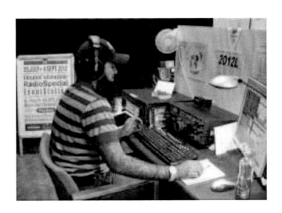

Saif A71AM operating 2O12L

Breaking records: The previous highest number of contacts made by a special event station in the UK was 47,791, made by the Cray Valley team at M2000A. The world record was 49,790 set by the DQ2006X team during the 2006 football World Cup. Both totals were exceeded by 2O12L. The team had set itself a target of 60,000 contacts, but went on to make 69,644 contacts in 220 DXCC entities during the 47 days of operation.

The Society was honoured to provide England's special event station for the 2012 Games. 2O12L made its final contacts as the Games' closing ceremony came to an end. This also marked the announcement that Dave G4BUO, Chris G0FDZ and I were finally retiring from organising major special event stations! A DVD of the event was produced by Fred G3SVK and sold many copies.

2013

January started well with five candidates successfully passing the Intermediate exam. New callsigns announced were: Giles 2E0HPY, John 2E0JKZ, Martin 2E0MWI and John 2E0TXJ. The success continued when the RSGB AFS SSB contest results were released: Cray Valley 'A' had won the contest for the second time in three years. Once again, arrangements were made to provide four competitive stations to maximise the number of contacts made by the 'A' team. Mark M0DXR operated from Dave G3RGS' QTH, Dave G4BUO operated from home with an antenna erected in a neighbouring field a few hours before the contest, I operated from Frank G4WNF's QTH, and Nobby G0VJG used the 9th Dartford Scout HQ. Mark took the plaudits, winning the contest outright with 326 contacts, and being awarded the *Lichfield Trophy*. Dave G4BUO was 8th with 309 contacts, I was 11th with 301 contacts. Nobby G0VJG's 286 contacts placed him 16th. The Society therefore won the *Flight Refuelling ARS Trophy* again. The 'B' team of Guy G0UKN, Ian M0UAT, Owen G4DFI and Martin M0MDR were placed 20th. Smudge G3GJW, Kevin M0KSJ, Karen G8JNZ, John M0JHB, Frank G0FDP and Cliff G4HSU made up the 'C' and 'D' teams. The Society was one of only two to field four teams. A great result!

The Society also entered the CW leg of the contest. Although the 'A' team of M0DXR, G4BUO, G0VJG and M0UAT could only achieve 25th place, Mark M0DXR once again topped the pile, taking first spot from Nigel G3TXF by seven QSOs. Dave G4BUO was 9th in the results.

Continuing the successful vein, a good team was available for a serious tilt at the SSB leg of the ARRL International DX Contest in February. M8C made 2,123 contacts and were placed 1st in England, 13th in Europe and 30th in the world in the Multi-Single High Power category. The team were Nobby G0VJG, Mark M0DXR, Richard G7GLW, Dave G4BUO, Ralph M0MYC, Toby 2E0TBO, Alan 2E0LAG, Giles 2E0HPY, Tim G4DBL, Garo G0PZA, Martin 2E0MWI, Graham M0PTD and myself.

182

Committee upheaval

Internal disputes between some committee members and officers of the Society over a period of a few months around the turn of the year and into 2013, came to a head at the beginning of February when allegations were reportedly made over the conduct of some officers of the Society. The Chairman referred the matter to the Vice Presidents and I to investigate. Unfortunately, the investigation had not been concluded before the Chairman resigned, followed by the Treasurer, Secretary and two committee members. As a result of the resignations, the investigation report was not concluded.

Following the resignations, the two remaining members of committee approached me to assume the role of Interim Chairman, which I accepted. An Interim Committee was formed, consisting of Frank G4WNF, Keith G4JED, Colin G3SPJ, Cliff G4HSU, Dave G3RGS, Dave G4BUO and Tom G1FAD. These members were happy to serve the Society for three months to manage day-to-day matters, make recommendations to the committee to be voted in at the Annual General Meeting, and plan for the Annual General Meeting itself. The Interim Committee met on several occasions to discuss a wide range of matters and to ensure members saw a seamless transition until a new committee was elected at the Annual General Meeting.

Construction initiatives

The annual Construction Contest was well supported with an array of varied projects, and was equally well attended by members. Chris G0FDZ hosted the event. Keith G4JED and Des G6WCX were the judges. Martin 2E0MWI received a 'commended' certificate for an audio amplifier Intermediate course project. He had fitted it neatly and securely into a box with necessary interfaces for the various connections, but importantly the connections had been carefully marked, making

Keith G4JED (left) presenting the Reigate Cup to Slim G4IPZ

the unit of use and identifiable to others. Commendations were also awarded to two projects submitted by Slim G4IPZ, who won the Reigate Cup for another of his entries, demonstrating incredible accuracy to solder small components and multi-leg Integrated Circuits. The evening showed the art of homebrew was still very much alive among Society members.

Following a RSGB 4 metre UKAC contest, Guy G0UKN flippantly suggested that the Society should undertake a transverter project to enable more members to operate on the 4m band. This was a popular idea. From several options, it was decided to customise the Elecraft XV50 transverter so it could be used on 4m. Investigations found that several amateurs had put the XV50 on 4m. Elecraft were contacted, terms discussed, and a favourable deal offered. 10 members built the transverter, and two built their own bespoke transverters. As a result, the Society now holds a successful 4m net.

New committee formed

The Annual General Meeting was well-attended and all the business expected of such a meeting was transacted. A new committee was formed under the chairmanship of Frank G4WNF. Slim G4IPZ was elected Vice Chairman, Lawrie G4FAA Secretary, and Colin G3SPJ returned to the Treasurer's role. Four committee members were elected - Cliff G4HSU, Dave G4BUO, Tom G1FAD and Martin 2E0MWI.

Meopham Windmill revisited and 2013 activity

GB6MW received its 6th outing, but once again the weather was not ideal for spending a weekend in a cold, draughty mill built in the 1800s! This led to many cups of hot tea and coffee and the odd trip to *The Kings Arms* across the road to warm up! The Society's new Acom linear amplifier had its first event outing and performed exceptionally well. Over the weekend, 428 contacts were made on HF – mainly on 40m. VHF was a more leisurely affair with 55 contacts, mainly on 2m FM. The mill open day on the Sunday had a stream of visitors enticed by bunting, welcoming signs and the Society's exhibition displays. There was even a military vehicle side-show as Martin 2E0MWI exhibited his radio-equipped Ghurkha Rifles Land Rover.

After 'retiring' as 'hounds', Guy G0UKN and Matthew 2E0MER became the 'fox' for the 2013 DF Hunt. Although the weather was favourable, there were too few 'hounds', although one team included Ben DL6RAI who was in the UK on business.
184

Only one team located the 'fox': Frank G3WMR and Dave G4YIB, who received the Reigate Cup at the post-DF Hunt social.

In the VHF NFD write-up in QUA, Colin G3SPJ expressed disappointment that only 12, mostly part-time, members had been involved. As a result, there had been empty seats at one or more of the operating positions for some of the contest. More support would have improved the overall result, which saw G3RCV/P achieve only 9th place in the Restricted section. Performances on 50 and 70MHz were the bedrock, being placed 7th and equal 4th respectively. These placings were helped by achieving ODX (best DX) of all 50MHz Restricted section entries for a QSO with SM5EDX at 1,371km, while on 70MHz a QSO with SW8YA (Greece) at 2,199km was ODX for any team taking part in VHF NFD. The team were also awarded 2nd best 'Centenary Points' and picked up an RSGB Centenary certificate.

A return visit to the home of 2O12L – the Royal Greenwich District Scouts Activity Centre in New Eltham – for the 2013 summer BBQ social coincided with one of the summer's hottest days. Over 30 members and guests enjoyed a fine array of BBQ ribs, sausages and chicken accompanied by an excellent range of salad. Dave G4NOW arranged the BBQ facilities, and Frank G4WNF and Nobby G0VJG provided the food. A raffle raised £30 for club funds.

After a brief absence, caused mainly by the 2O12L activity, a Cray Valley team again headed westwards to the Isles of Scilly at the end of July for the RSGB IOTA contest. The 2013 team was Nobby G0VJG, Richard G7GLW, Garo G0PZA, Guy G0UKN, Ian M0UAT and Chris G0FDZ. They had to contend with two rough crossings on the *Scillonian,* and being placed 5th in the 24 hour DXpedition category with 2,290 contacts and a score of 898k. Outside the contest, the team were in demand to provide a contact with EU-011.

Following an enthralling talk on the WW2 German Enigma coding machines by Martin G3ZAY, a visit was arranged to the once top-secret Bletchley Park complex, home of the WW2 code-breakers and now a museum. 20 members and guests were treated to a personalised guided tour by Martin, which began with an introductory talk, followed by a walk around the site, stopping off at various external points of interest. After the guided tour, members explored some of the unvisited huts, including, at Block H, the home of The National Museum of Computing, which, as in 1944, housed the Colossus machine. Several members also looked around the mansion and

Martin G3ZAY with members and guests at the start of the guided tour at Bletchley Park

were rewarded by a short impromptu talk by the site manager about the impressive internal décor.

The Society was invited back to Woodlands Farm at Shooters Hill for the 'Apple Day Fete' in October. GB4WFT was aired again. Essentially a fun day, even though the weather was particularly unkind, the opportunity was taken to demonstrate and talk about the many aspects of amateur radio.

The Society organised its 9th Foundation Course in November. Six of the seven candidates were successful. Having completed the course, all were granted a free year's membership. November was also a busy month on the surplus equipment front, with a Surplus Sale and three sales tables at both Kempton Park and the Coulsdon ATS Bazaar. The three sales brought a net profit of £561, which added to the sale of other surplus equipment during the year, meant the Society's equipment fund had grown by £1,140 in 2013 in readiness for future equipment purchases.

G3GVV SK

Tim G3GVV passed away peacefully in December aged 95. He was a past RSGB President in 1972, and worked tirelessly in international affairs for the RSGB, attending a number of IARU and ITU conferences. He established excellent relations with numerous ham radio representatives around the world and chaired the RSGB IARU Committee for many years. Tim joined the Society for M2000A when we needed people to help with the schools programme. Tim was a natural choice. He dedicated a lot of time to giving introductory talks to groups of visiting schoolchildren, and having established a relationship with the Society, remained a member until failing eyesight meant he could no longer read QUA.

Frank G4WNF organised the Christmas meal at a new venue, *The Lord Kitchener* in Welling. It was a popular decision. Not so popular was the decision to hold the Christmas drinks evening at *The Royal Oak* in Bexleyheath. Attendance was poor, possibly due to the location and being perceived to be poorly served by public transport.

2014

The Society's first activity of the year was in the RSGB AFS CW contest. A team of Mark M0DXR (operating from Dave G3RGS's QTH), Dave G4BUO, Garo G0PZA and Nobby G0VJG were placed 15th, but Mark led the way with 276 contacts, winning the *Marconi Trophy*. In the RSGB AFS SSB contest, the 'A' team of Mark M0DXR, Nobby G0VJG, Dave G4BUO and myself could only manage 4th place. The 'B' team of Guy G0UKN, Owen G4DFI, Ian M0UAT and Kevin M0KSJ were 24th. Martin M0MDR and Cliff G4HSU formed the 'C' team.

Keith G4JED organised a successful 'planning meeting' to discuss the Society's programme and activities for the year. Feedback obtained suggested members were particularly keen to see the Meopham Windmill event and the summer BBQ repeated, together with an entry into VHF NFD. They offered suggestions for several organised visits and special event stations, and that the Christmas drinks evening should be at the *Park Tavern* in Eltham.

Following success in the Advanced exam for the Full Licence, four new M0 callsigns were announced - Giles M0TGV, Martin M0MWI, Richard M0NAI and Steve M0HOH. Later in the year, Toby 2E0TBO and Alan M6DYI, who had both obtained licences at earlier Society-run courses, became M0TBS and M0HPZ respectively.

M8C in the CQWPX SSB contest
The Society had not entered this contest seriously since 2006, when a score of 4.1M took the Multi-operator, Multi-single high power record for England. Nobby G0VJG assembled a 13 man team in 2014 to try to better that score. Operating from the 9th Dartford Scouts HQ, the team succeeded making 3,072 contacts in

the 48 hours for a score of 7.98M. The score took 1st place in England, and was 18th in Europe and 31st in the world. It was also the 11th best score in that category to be made by an English team (up to and including 2015). Non-contesters were able to use a second

Garo G0PZA operating M8C in CQWPX

station to work DX on 12 and 17 metres. The successful outcome was undoubtedly helped by a Saturday barbeque and Sunday roast dinner!

Annual General Meeting

A new committee was elected at the Annual General Meeting with Frank G4WNF elected as Chairman for a second year, with Colin G3SPJ as Vice-Chairman for 2014/15. Martin M0MDR became Treasurer and Rich G7GLW took over the role of Secretary. Cliff G4HSU, Dave G4BUO and Tom G1FAD were elected as committee members. Keith G4JED received the Founders' Cup in recognition of his outstanding work as programme secretary over five years, and Chris G0FDZ was presented with only the sixth ever Cray Valley Diamond Award for obtaining 100 points in contacting members, club stations and special event stations organised by the Society. This is the ultimate Society award and requires much dedication over a long period of time. Indeed, it took Chris 13 years to make the required number of contacts.

Four organised visits and another trip to the windmill

Members' comments at the Annual General Meeting were heeded as four organised visits were arranged during the year. The first three were the Secret Nuclear Bunker at Kelvedon Hatch in Essex, the BT Kingsway Tunnels complex close to Chancery Lane in central London, and the Kirkaldy Testing Museum at Southwark. The level of participation in all three visits was good. Facilitated by David M0XBO, the visit to the BT tunnels was a unique opportunity to visit a tunnel complex that was built during the last war and adapted for cold-war use, and the Kirkaldy Museum gave an opportunity to see testing equipment from the 19th and 20th century in its original building. The fourth visit of the year was to
188

Crossness Museum for one of their open days. The scale and magnificence of the beam engines in the Crossness building was something to behold, and this became the venue for future special event stations during Museums on the Air weekend.

L-R: Frank G3WMR, Kevin M0KSJ, Richard G8ITB, Alan G0HIQ and David G6NRH at the Secret Nuclear Bunker

Once again the weather played a major part in the Mills on the Air weekend at Meopham windmill. With gusty winds and cool temperatures, the team were grateful for Frank G0FDP's hot tea and coffee and Guy G0UKN's egg/bacon/sausage rolls and a couple of fan heaters! Undeterred, members used the Society's Icom IC-756PROIII and Yaesu FL2100Z linear amplifier on HF. GB6MW made 320 contacts, mainly on 40m, but 15m provided a few stations from Japan, VP8LP (Falkland Is), and Mississippi member, Mike WM5DX. On 2m, 63 stations were contacted. The Mill Trust, as always, were very welcoming and appreciative of the Society's continued support for the mill – especially welcome in times of financial cut backs, enforced delays to maintenance and a rather uncertain future (in terms of opening to the public) for many mills.

2014 club project

For the second year running a construction project was organised by Guy G0UKN. A decision was taken to build the Radio-Kits digital SWR meter. The project was undertaken by members as a group build, to encourage non-constructors to take part. All components were of standard size and no surface mount soldering was involved. Chris G0FDZ and Guy built prototype meters and their experience of building the kit helped others to complete the project successfully.

A disappointing turnout for the DF Hunt and VHF NFD

Only seven members took part in the 2014 DF Hunt, the timing of which coincided with England's World Cup football match with Uruguay. Neither team discovered the 'fox' – Frank G3WMR and Dave G4YIB. Interest in the event had been on the wane for a few years, and this poor showing led to the committee deciding to cancel the event in 2015, in the hope of a revitalised event in 2016.

A full 4-band entry in VHF NFD was not possible as several key members were unavailable. The Society was represented in the results listings as Lawrie G4FAA organised a small team to operate G3RCV/P in the Restricted section from the Kent County Show Ground at Detling, near Maidstone. Using 100 watts into a hand-rotated 9-element yagi, G3RCV/P made 121 contacts, but languished in an unfamiliar 16th place. However, judging by the RSGB Contest Committee write-up, the Society was not the only group with difficulties as it spoke of mediocrity being a depressing word, but that it probably embraced the weather, the operating conditions and the activity level.

New roll-up banners purchased...ready for GB2CM!

Thanks to Kevin M0KSJ, the Society obtained two new roll-up banners to enhance publicity about the Society and our exam courses at public-facing events. They were first used at GB2CM, which is a great lead into the Society's first special event station from Crossness Museum for the Museums on the Air weekend.

The Crossness Pumping Station at Erith/Thamesmead was built by Sir Joseph Bazalgette as part of Victorian London's urgently needed main sewerage system. It was officially opened by the Prince of Wales in April 1865. The Beam Engine House is a Grade 1 Listed industrial building constructed in the Romanesque style featuring some of the most spectacular ornamental Victorian cast ironwork to be found. It also contains the four original pumping engines (although the cylinders were upgraded in 1901). These are possibly the largest remaining rotative beam engines in the world, with 52-ton flywheels and 47-ton beams.

The Crossness Engines Trust was set up in 1987 to restore the installation, which represents a unique part of Britain's industrial heritage and an outstanding example of Victorian engineering – a large part of the restoration work being carried out by an unpaid volunteer workforce. As part of its fundraising 'The Beam Engine Trust' hold public open days, known as 'Steaming Days', during which clubs and societies display information about their respective interests and hobbies.

A perfect summer's day attracted over 500 visitors to the event. Apart from the GB2CM display, the Erith Model Railway Society and the South East London Meccano Club provided impressive displays. Other exhibits included large scale live steam locomotives, exquisite boats, lots of books, kits and much more. Outside, stationary engines and model traction engines tooted and steamed amongst the admiring onlookers. From an amateur radio perspective, the day was a huge success: it was well-supported by members, the exhibition was popular with visitors, and 232 contacts were made.

Equipment loan policy introduced

The committee introduced a loan policy, whereby certain items of Society equipment were designated as being available for loan to members on a short-term basis. Enquiries and applications were to be made to the Equipment Manager or, in his absence, a committee member. Applications would be assessed before approval was given. The fact that the list of loanable items was quite narrow caused a re-think, with the policy subsequently revised and re-issued in 2015.

IOTA DXpedition to Les Minquiers

A five man team, comprising Chris G0FDZ, Giles M0TGV, Toby 2E0TBO, Nick G4FAL (who joined the Society specifically to be part of the team), and Nobby G0VJG mounted a DXpedition to the uninhabited island of Maitresse in the Les Minquiers group (located between Jersey and St Malo) in July for the 2014 IOTA contest. The islands had their own unique IOTA reference – EU-099. Like most

Nick G4FAL and Toby 2E0TBO operating MJ8C from Les Minquiers

DXpeditions to rare DXCC countries, the trip was fraught with logistical challenges: the team had to be taken to the reef by the Jersey Coastguard, had to take around 200 litres of fresh water, food for five people for seven days, fuel for two generators, medical supplies, as well as the amateur radio kit of transceivers, antennas, coax, tools, etc. Due to the rarity value of EU-099, the team made over 6,000 contacts outside the IOTA contest using the GJ3RCV callsign. Another 1,693 contacts were made using MJ8C during the contest. The team were the leading British Isles DXpedition in the Restricted category with 3.69M points.

G0FDU and G3JJZ SK

It was with great shock that members were advised of Ray G0FDU's passing. Ray joined the Society in 1988 and had been a committee member, and was part of the M2000A organising committee in 1999 with responsibility for datacomms. Prior to his death, Ray was a very active member of the Darenth Valley Radio Society (G0KDV).

Dave G3JJZ passed away at the age of 77 at around the same time. He obtained his licence at the age of 16 and was an avid CW operator. He was a member of the Society from the late 1990s for about 10 years. He had been a committee member, and was Vice Chairman in 2005. Dave enjoyed the operating side of the hobby, entering the RSGB AFS contests and taking part – and winning – the annual DF Hunt on several occasions. Dave was also responsible for a number of members acquiring their amateur radio licences, including Simon (now 2E0CVN) and the Society's President!

Further special events

GB2HTL at Holy Trinity Church, Lamorbey, Sidcup, was organised by Phil G8OPA for the Churches and Chapels on the Air weekend, an annual event run by the World Association of Christian Radio Amateurs and Listeners. The event was a success on three fronts – a successful PR exercise, over 100 stations contacted in a six hour presence on 15, 40 and 2m (including 16 WACRAL stations), and also attracting a new member (Julia G4YMF), and her husband as a candidate for the next Foundation course.

Owen G4DFI, the Society's QSL Manager, operating GB2HTL

Instead of activating GB2CM again, members manned a static display at a second Crossness 'Steaming Day' and local history fair. The display, which was extremely

192

popular, featured local radio history in the guise of memorabilia from local companies Kolster-Brandes Ltd, KW Electronics and Burndept/Vidor. The addition of medium wave '30s and '40s music from classic domestic radios was a crowd pleaser. Visitors included former KW Electronics employees Chris G8GKC, John G8JAD and John G4MGY who had not seen each other for a number of years. They shared many reminiscences about the company – even down to pointing out some of their work on the kit on display!

A team returned in October for the last 'steaming day' of the year for *'Prince Consort'*, one of the huge beam engines. Once again, the static display embraced the theme of the Open Day with a mix of local radio history and displays of modern amateur radio.

Cray Valley's major input into '2SZ'

Although not a Cray Valley event, eight members had a major influence on the special event station established at Mill Hill to mark the 90th anniversary of 18 year old Cecil Goyder 2SZ's first ever radio contact from the UK to New Zealand in 1924, which proved that global communication by radio was possible.

The Society, through Dave G4BUO, was approached for assistance due to its track record in organising and running major special event stations. It was particularly relevant that the Society should be asked as this history of the Society has already featured the 50th anniversary of the contact, which was marked by members in October 1974 using the callsign GB2SZ. Ofcom re-issued 2SZ and over 20,000 contacts were made during the week long activity.

Mast installed at meeting venue

The Society had kept George G3BNE's Altron mast following his death with a view to installing it at the meeting venue. The opportunity arose, through the good offices of Frank G4WNF, for the mast to be erected. Later, a 6/2/70cms vertical and a 40/80m dipole (donated by Owen G4DFI) were added. Having antennas permanently installed would mean a time saving on exam course days as it would no longer be necessary to set up or take down the antennas on the day. However, 'local issues' were to lead to a re-think. More on this as we move forward to 2015.

The end of a busy year

With no serious activity planned for either the CQWW SSB or CW contests, the committee agreed Tim G4DBL's request to use M8C in both. Although they were not winning entries, it enabled M8C to appear in the results listings. Tim made 433 contacts in six hours in the Single Operator All Band (SOAB) category of the SSB contest, and 107 contacts in six hours in the 10m Single Band Low Power category. Both entries earned certificates.

The Society's participation in the RSGB 2m UKAC contests had been poor in 2014, with the majority of contributions from just five members. The winner of the G3BNE Cup was Guy G0UKN. Malcolm G8MCA scored most heavily, but as winner in 2013, he was ineligible to win it for a second consecutive year. Certificates for the band winners were awarded at the discretion of the committee, taking account the level of participation.

The Society ran its ninth Foundation course. Eight of the 11 candidates passed – three with maximum marks. The eight new callsigns were announced as – Akimoto M6FBN, Jon M6HSN, Dorn M6YMJ, David M6EYC, Harry M6HGC and Adnan M6XML; seven signed up for the January 2015 Intermediate course.

The two Christmas festive meetings were well-supported; 38 members and guests attended the Christmas meal at the *Lord Kitchener* in Welling, while 20 members attended the Christmas drinks evening at the *Park Tavern* in Eltham.

2015

The year began with all nine candidates passing the Intermediate exam, and two successfully re-sitting the Foundation exam. Members had real pleasure witnessing the sheer delight of one severely disabled candidate when it was announced that he had passed at the third attempt. The success was largely due to the efforts of Dave G3RGS who personally tutored and mentored John for

The successful crop of 2015 exam candidates

six months. As a result of the two successful courses, QUA carried advice of the new callsigns – Akimoto 2E0FBH, Jon 2E0HSN, Steve 2E0HOU, Dorn 2E0ORN, David M6EYC, Harry 2E0HGJ Richard 2E0RJX, Adnan 2E0LSY, Jerry 2E0JCJ, Arnold 2E0LTU, John M6WCQ and Dave M6FDG.

AFS contests
Due to the sheer number of entrants in previous years and the loss of the non-contest segment of 80m (3.650 – 3.700MHz), the RSGB Contest Committee decided, as an experiment, to allow 40m to also be used. Strategy therefore came into the contest, with decisions required as to which band to start on. Those starting on 40m fared well, with several members accumulating 100 contacts in less than 35 minutes. Seven members made over 240 contacts. With the overall positions calculated using normalisation (the winner of each band scored 1,000 points and everyone else received a pro-rata score), it was impossible to predict the outcome.

Although the Society did not pick up the main prize, Mark M0DXR made equal most contacts – 468, and I made most contacts on 40m – 314. The 'A' team of Mark M0DXR, Dave G4BUO, Nobby G0VJG and I were 3rd behind the Bristol Contest Group and Cambridge Hams 'A', but it was close as there were only 13 contacts between 1st and 3rd. The 'B' team of Malcolm G8MCA, Garo G0PZA, Giles M0TGV and Owen G4DFI were 18th. Ian M0UAT, Frank G4WNF, John M0JHB, Cliff G4HSU and Martin M0MDR also represented the Society. Mark M0DXR was the only member to represent the Society in AFS CW, making 219 contacts.

K1N weekend
Navassa Island, No.1 on the DXCC Wanted List, was activated in February. As the pile-ups would be huge, Nobby G0VJG arranged for a competitive station to be set up at the 9th Dartford Scouts HQ to give members the opportunity of working 'a rare one'. With a 2-element quad at 70' and FT-2000 and FTDX3000 transceivers and an Acom amplifier, 12 members had contacts with K1N. The weekend was a great success. As well as K1N, 133 DX contacts were made, and there was a good supply of food – hot bacon, sausage and egg rolls and burgers, tea and coffee. The members who attended each donated financially to offset the cost of the hall hire and the food.

In March, M8C was aired by Tim G4DBL in the ARRL SSB DX contest, and myself in the Russian DX Contest. Tim made 85 contacts in a 10m only entry. I made 525 contacts and achieved 1st place in England and 33rd in Europe in the SOAB SSB category.

Annual construction contest

The evening was well-attended featuring 20 entries from 11 Society members. In addition, although not entered into the contest, Chris G0FDZ showed the GB3UHF 70cm beacon which had been built by Chris and other members of the beacon team. It was fully functioning with a GPS synchronised clock, but was awaiting the appropriate permission to be installed.

The contest judges were Doug G0FAS and Dave G3RGS. The construction project led by Guy G0UKN featured strongly with six digital SWR meter entries, each one built to a very high standard. Two entries, from first time constructors Dorn 2E0ORN and Harry 2E0HGJ, were based on Intermediate exam construction projects. There were other entries from John M0JHB, Malcolm G8MCA and Slim G4IPZ. Overall, entries were judged to all be of an exceptional standard. After considerable deliberation the judges concluded that the auto-switched 4/6m PA constructed by Malcom G8MCA just edged Slim's Re-Flow Oven project. It was also considered that the SWR meter entered by Frank G0FDP was the best entry from the club project group; the audio amplifier made by Harry 2E0HGJ received special mention for its high standard of workmanship.

Annual General Meeting

Over 30 members attended the 2015 Annual General Meeting. Frank G4WNF conducted the meeting, but he, Colin G3SPJ and Tom G1FAD did not seek re-election for 2015/16 after giving much time and effort on behalf of the Society. A full committee was elected, with Dave G4BUO as the new Chairman and Rich G7GLW as Vice Chairman. Giles M0TGV joined the committee. Martin M0MDR and I jointly filled the role of programme secretary until April 2016, following the hiatus that arose when Keith G4JED vacated the position in 2014.

Martin M0MDR noted the financial stability of the Society in his Treasurer's report and Frank G4WNF gave a good account of the Society's activities during his tenure as Chairman. Mention was made of the Society's 70th anniversary in October 2016 and that the committee were scheduled to discuss the logistics of a full programme of events.

G3RCV shack developments (1)

This was an ongoing matter through the year. In February, a local resident apparently advised the Royal Borough of Greenwich's planning department about the mast at the Society's meeting venue. The Council's enforcement team contacted the Society with a view to it applying for retrospective planning permission. Although not invited to remove the mast, it was decided to do so and submit an application for a new 30' free-standing Tennamast in an almost identical position to support an A3S tri-bander and a 2m yagi. The committee believed the option of a stronger mast at a greater height represented a better approach to the ultimate aim of having an operational shack at the Scout hall.

Kevin M0KSJ worked on the necessary plans and drawings to accompany the planning application, Guy G0UKN took the proposal to the Scouts committee for approval, and I began work on the planning application submission. Scout approval was obtained, and I met the Council's planning officer to obtain 'pre-planning advice'. The full planning application was submitted in April, and planning approval granted in June. Read about the next phase as we progress through 2015.

Cray Valley moves into the 21st century!

Martin M0MDR set up a Twitter account for the Society. Follow Cray Valley Radio Society at: **@G3RCV #CrayValleyRS**

The Society also set up a Facebook page: '**The Cray Valley Radio Society Facebook Page**'. The link is **https://www.facebook.com/groups/660383353989487/**

Meopham Windmill...take 8!

GB6MW was as popular as ever for members, visitors and the Mill Trust. A permanent reminder had been added to the mill's exhibition area since the 2014 operation. Following the contact with Bob VP8LP in 2014, both QSL cards and information about the mill event and amateur radio's part in the Falklands invasion had been mounted in a frame and given to the Trust to add to their displays highlighting events associated with the mill over the years. Amateur radio was therefore added to the mill's list of plane crashes, coroner's courts, suicides and subsequent haunting!

The weekend event had grown since the first airing of GB6MW, such that 52 UK and 30 European mills took part in 2015. Operation was on VHF (ground floor) and HF (first floor). In around 12 hours of operating over 400 SSB contacts were made (mainly on 40m) with a few on 15m, including 33 UK mills. A more leisurely pace was enjoyed on VHF, but the station was able to make two DX tropospheric contacts into Spain and southern France. The weekend activity offered some of the Society's newer members the opportunity to experience operating and logging at a special event station.

Inaugural Gulf of Mexico awards

To foster closer liaison between the Society and the Great Southern DX Association (K5GDX) in Mississippi, dual member Mike WM5DX devised *'The Gulf of Mexico Award'* as an annual competition for contacting members of both societies. Mike was present at the December meeting to present 2015 plaques to Frank G4WNF, Dave G4BUO and myself.

150 years of Crossness pumping station

As 2015 marked the 150th anniversary of the opening of the Crossness Pumping Station, the Society operated a special event station from the main hall during the last 'Steaming Day' in June. Ofcom approved the application for the special call-sign GB150CM, which was used on the HF bands. A static information display was also provided. Much of the preparatory work was undertaken in advance, so only the antennas needed erecting on the day.

The special callsign was popular on SSB and CW, but local QRN (traced eventually to a laptop PSU) prevented some weaker stations making it into the log. However, over 400 stations were contacted in around six hours operating. A sizeable Cray Valley team enjoyed a successful day explaining amateur

Guy G0UKN operating at GB150CM

198

radio to visitors, running pile-ups, and discussing the intricacies of the Crossness Pumping House to radio amateurs over the air. Jan M3UJR and Matthew 2E0MER, both in Victorian dress, helped run that most valuable of enterprises – The Café!

A second activation of the special 150th anniversary callsign took place in August. The set-up was similar to the June event - Icom IC-7400, Acom 1000 linear amplifier and Palstar BT1500A, balanced ATU fed with 450 ohm ladder line to an inverted-V multi-band doublet at 28'. A total of 133 contacts were made, mainly on 40m SSB, but a few PSK31 contacts were made using Fldigi software. A few VHF contacts were also made. The static display was again popular, this time including the local radio history exhibits and the '30s and '40s background music used at the 2014 GB2CM event.

The GB2CM and GB150CM events certainly ticked all the boxes for a special event station: they raised public awareness of amateur radio, importantly with good attendances; raised awareness of the Society both locally – with around 800 visitors and almost 800 radio contacts; provided good opportunities for members to operate a variety of modes; and provided a good social occasion for members with good inside facilities in a fascinating location.

Summer BBQ

Dave M6FDG organised and ran a very successful BBQ. Dave had paid for a new BBQ to be made by Frank G4WNF's company and donated it to the Society for future BBQ-style events by way of thanks for helping him through the Foundation exam. The BBQ itself was held in the grounds of our meeting venue from 4pm. The sun shone and 37 members and guests ate a vast amount of burgers, steaks, ribs, chicken and sausages. Everyone even had room for gateaux, strawberry tart and chocolate cake! A free (yes, free!) raffle was organised, with Clare purchasing the non-radio prizes and a food hamper, won by Frank G0FDP. With skies threatening after a hot and humid day, the afternoon/evening event closed around 8.15pm.

Unsuccessful joint VHF NFD venture

The Society was approached by the Maidstone YMCA club with a view to entering a combined team in VHF NFD. The committee approved the plan as they believed it would be a way of enabling those members who wanted to take part to do so. However, it was not a successful venture.

There was no QUA report from a Cray Valley perspective, apart from a comment that the best part of the weekend was Dave M6FDG's BBQ and the views across the English Channel! The report posted on the Maidstone club's website made depressing reading. It spoke of storms and winds in the Channel which damaged the 2m tent, knocked the 2m antennas out of line, totally wrote off the 70cm tent and flattened sleeping tents. That was before the contest! During the contest there were issues with the coax to the 70cm antenna which led to be mast being lowered in darkness with dramatic results; 'deaf' 2m and 6m stations; a 4m station that ceased operation due to static rain; and a 23cm station that was abandoned before the contest began! As a result of the weekend's events, the committee resolved that it was unlikely to sanction any further joint contesting ventures with the Maidstone YMCA club in the near future.

DXpedition to St. Kilda

After IOTA DXpeditions to the Isles of Scilly and Les Minquiers, the team decided it would activate St. Kilda, a small remote group of islands about 40 miles off the Scottish west coast for the 2015 IOTA contest. Again, the pulling power was another rare IOTA reference – EU-059. I cannot do justice to the professionalism, resilience and dedication of the team in these few paragraphs.

However, the team of Nobby G0VJG, Dave G4BUO, Chris G0FDZ, Giles M0TGV, Tony G2NF and Justin G4TSH knew in advance that the weather in July could be challenging, and they were not disappointed, as they experienced a Force 8 gale with winds gusting in excess of 60MPH on their second night on the island. Masts were lowered, but not before one had been bent by the wind, and a particularly strong gust bent three of the six support poles of the cooking tent. That was

lowered and sandbags put on top to prevent it blowing away. Three sleeping tents were also damaged.

The St Kilda team: Justin G4TSH, G0FDZ, Dave G4BUO, Tony G2NF, M0TGV and Nobby G0VJG

So instead of running pile-ups on the WARC bands, the next morning was spent sorting out the mess, repairing the tents and re-erecting the masts and antennas. The weather was not particularly kind for the remainder of the team's stay, but there was a brighter interlude which enabled the team to climb the highest point on the island of Hirta, to activate the SOTA summit (SI-098) of Conachair at 430 metres. Sixty SSB contacts using 25w from a Yaesu FT-897 and 70 CW contacts using 10w from an Elecraft KX3 were made from the summit.

Outside the contest, the team used GM3RCV/P. Several nets were arranged for members to contact St.Kilda, and 17 different members were worked during the contest. In the contest as MM8C, the team used a Funcube Dongle Pro Plus and Skimmer software to aid their search for multipliers on CW, but radio conditions were bad! The team made only 11 contacts on 15m and worked a solitary Spanish station on 10m, and were disappointed making only 992 contacts. This gave a checked total score of 2.16M and 2nd place in Scotland and 11th in the Low Power Multi-Single DXpedition category.

The journey home was delayed by bad weather, ferry cancellations and overnight roadworks, with the team eventually arriving home more than a day late. Having once again raised the Cray Valley profile, the team were awestruck by the island, and humbled learning of its history.

G3RCV shack developments (2)

The Tennamast 10.5m mast was delivered to the meeting venue in late July. Dave G4BUO, Kevin M0KSJ, Chris G0FDZ and Frank G4WNF installed the mast base in August and the Scouts' builder (called Bob, incidentally!) poured the concrete once the base post had been set.

An antenna party comprising Dave G4BUO, Dave M6FDG, Colin G3SPJ, Chris G0FDZ and Frank G0FDP gathered at the Society meeting place in early September to install the mast, antennas and cable trunking. Colin G3SPJ donated the 2m yagi, and the Cushcraft A3S tribander and rotator had been given to the

The new Tennamast and antennas at the Society's meeting venue

Society by G3GVV's family. The installation was carried out in accordance with the approved planning application.

Fred G3SVK was invited to make the inaugural contact before his 'Top tips for working DX' presentation in September. HZ1SK in Saudi Arabia was that first contact, with 59's exchanged on 20m SSB.

Fred G3SVK in contact with HZ1SK

There was still work required to add a 40m dipole to the mast, and in the *'Leaders' Room'* where the shack was to be located. Guy G0UKN donated a desk, which he was to turn into a lockable cabinet housing the G(X)3RCV station, and because he was also on the Scout committee, obtained their agreement for the free use of the room, kitchen and lavatories on one day a week. However, details of the arrangements for using the station were delayed due to local interference issues.

G3YOU SK
It was with deep regret that members were advised of the passing of John G3YOU in September. He had been poorly, but his death was unexpected. John had been a member of Cray Valley in recent years, but had been a big part of the West Kent ARS for very many years. John was a gentleman and always ready to help where he could to develop individuals at both societies. In 2014 he contacted W1AW, the ARRL centenary station, in all the US states; his achievement certificate arrived a few weeks before his death. Members of both societies were present at his funeral service.

SSB Field Day and the 2m Open
The Society took part in both contests – G3RCV/P in SSB field day and M8C in the RSGB 2m Trophy contest. HF conditions were poor in the early part of the contest, but improved with good runs on 40, 80 and 20m through until 02.00z. The nine contacts made on 10m were the most made by any station in the

contest! After checking, G3RCV/P was 3rd behind the Bristol Contest Group and the East Notts Contest Group. Both Steve M0SCJ and Dave M6FDG made their first contest contacts; it was hoped the experience might lead them to participate in future HF contests.

With only 50w to a 9-element yagi at 4m, a site only 50' ASL, and breakthrough from the SSB field day station, the M8C entry into the 2m Trophy contest was never going to be competitive.

GB0RWC for the Rugby World Cup

The Society was invited to run GB0RWC, the special event to celebrate the 8th Rugby Union World Cup for four days in October. Unfortunately, the operating plan was seriously affected by the local interference issue which saw the noise level at s9. 40m operation was not affected so much due to the strength of the majority of the signals, but the operators found it almost impossible to pull the less well-equipped stations out of the noise. Poor band conditions did not help either.

Richard G7GLW operating at GB0RWC with Richard G8ITB

Day 1 saw over 200 stations contacted, mainly on 40m but the best was perhaps a Dutch station in JO32 which showed that the new 2m set-up of 100w to an 11 element yagi appeared to be working well. 150 contacts were made on day two, with the best DX into the Middle East on 20m. More contacts were made early on day three, but after some detection work to try to better identify the noise source, a decision was taken to cease operation after lunch.

JOTA from Severndroog Castle

Severndroog Castle is an 18th century gothic folly located in Oxleas Woods close to the top of Shooters Hill in the Royal Borough of Greenwich. Being one of the highest points in London, the viewing platform at the top of an 86 stair spiral staircase provided a 360 degree view of London and seven counties. It also provided the opportunity to hang nested dipoles for 40, 20 and 10m from the top as a sloper!

The GB2RE station, exhibition area and DF Hunt for the Scouts was arranged largely by Guy G0UKN and Matthew 2E0MER. The event also featured a Skype webcam enabling the Scouts to take part in JOTI (Jamboree on the Internet). The station, on the second floor of the Castle, comprised the Society's IC-7400 transceiver and Guy G0UKN's IC2KL linear amplifier. There was also an SWL station on the first floor together with some promotional information as well as

introductory talks to amateur radio and Morse appreciation sessions. In the woods surrounding the Castle, a DF hunt was organised by Matthew 2E0MER and Richard G8ITB, for groups of scouts to locate a hidden micro transmitter.

Scouts waiting to take part in the DF Hunt

GB2RE was on the air for nine hours. In that time, Guy and I were able to let over 80 Scouts and leaders pass messages to radio amateurs and Scouts at other JOTA events in the UK and around Europe. Best DX of the day was the first contact into Australia, but another notable contact was with the Hampshire Winter Camp staffed by old friends from 2O12L, Brian G0UKB and Liz M0ACL.

Moving to the end of the year

The main December meeting saw the Society host a quiz with the Loughton & Epping Forest ARS. It was a less than serious event which the visitors won by 46 points to 40. The 11th Foundation course was a complete success with all five candidates obtaining a pass. As a result, new callsigns were announced as – David M6WHF, Glen M6GSN, Remus M6HTV (who was also successful in the Intermediate exam at a different venue), Simo M6AET and Simon M6SLZ (who was also awarded the 'Sandell-Whitehead' trophy). Additionally, Adnan 2E0LSY passed the December Full exam and obtained the callsign M0SLM; he had completed the route from Foundation to Full licence in just over one year.

To end the year, the festive meetings were well supported: 38 members and guests attended the Christmas meal at the *Lord Kitchener* in Welling, while 18 were present at the Christmas drinks and free buffet evening at the *Park Tavern* in Eltham.

70 years later...2016!

Changes were announced to the Society's successful training arrangements. Kevin M0KSJ assumed the role of Training Co-ordinator, with responsibility for publicity and the registration of course candidates. Chris G0FDZ continued to be Lead Instructor and retained responsibility for all aspects of the training and organisation of each course. The changes were believed to be required to address changes to the training situation in the UK, which had seen the emergence of on-line course providers. This development required the Society to be much more pro-active in securing candidates for its future courses.

11 members took part in the RSGB AFS SSB contest. It was again a two band affair (40 and 80m). Martin M0MDR aired G3RCV from the club shack for its first competitive appearance. The 'A' team of Mark M0DXR, Giles M0TGV (with Nobby G0VJG), Martin at G3RCV and I were 2nd

Martin M0MDR with Shack Manager Trevor M1TAD (left)

behind the Bristol Contest Group. The 'B' team of Garo G0PZA, Owen G4DFI, Frank G4WNF and Richard G8ITB were 25th. John G0JHB, Cliff G4HSU and Lawrie G4FAA made up the 'C' team. In AFS CW, Mark M0DXR was a team of one, but achieved 1st place and was awarded the *Marconi Trophy*.

70th anniversary outline

As the Society moved closer to its 70th anniversary, a sub-committee of me, Chris G0FDZ and Owen G4DFI was formed to plan the month of activity, culminating in an anniversary party. QUA released an outline of activities planned for October 2016:

- A talk about the Society in the 40s, 50s and 60s
- A special event callsign for use all month
- A special anniversary logo
- A 'Platinum' award (an engraved plaque) for contacting club members
- An activity week to attract applications for the 'Platinum' award
- An Anniversary party for current and ex-members
- 70th anniversary merchandise

It was also planned to release a 'Project 70' USB stick containing memorabilia from the Society's 70 years and this history book towards the end of the year.

The early part of the year

Following their success in the Advanced exam for the Full amateur licence, nine new callsigns were announced – Harry M0VMF, Akimoto M0HZB, Richard M0RMI, Jon M0OVA, Arnold M0PZD, Paul M0WPG, Remus M0UKL, Dorn M0IAO and Steve M0OSY.

Society members manned a static exhibition at the St. Mary's (Eltham) Community Complex Open Day in February, which provided a further opportunity to showcase the Society and amateur radio in general. The professional looking display was complimented by the organisers, but visitor numbers were poor.

There were four sections to the year's construction contest – 'Novice', 'General', 'Professional', and 'Not Judged'. There were three entries in the 'Novice' section, two in the 'General' section, and two in the 'Not Judged' section. Colin G3SPJ, Frank G4WNF and Dave G8ZZK judged the competition and awarded the Reigate

Cup to Richard G8ITB for two boxed and engraved interface converters for use with his Icom transceiver.

Shack news

Dave M6FDG and Colin G3SPJ carried out tests using a CAT scanner to trace the power cables in and around the scout hall, but the local interference issue which had prevented regular activity from the G3RCV shack was no longer evident. Trevor M1TAD, the newly appointed shack manager, was then able to open the shack to members on a regular basis. On one occasion, a few members were able to contact Society members Giles M0TGV and Nobby G0VJG on their DXpedition to Botswana (A25UK).

As the free use of the Leaders' Room on one day a week had been negotiated with the Scout committee, members repaid that generosity by painting the outside of the meeting venue over two days; the fresh coat of paint transformed the look of the building completely.

Annual General Meeting ballot required

Members at the Annual General Meeting elected Dave G4BUO as Chairman for a second term. Richard G7GLW was confirmed as Vice Chairman, Martin M0MDR as Treasurer and Karen G8JNZ as Secretary. Unusually, a ballot was required to elect the committee after five nominations were received. As a result, Trevor M1TAD joined the committee and Cliff G4HSU became a 'back-bencher', after many years on committee.

It's mill time...for the ninth time!

Once again, the weather was not ideal for the ninth activation of GB6MW from Meopham Windmill. The G0UKN breakfasts did a roaring trade, as did the G0FDP tea shop! Band conditions were also poor, with a lack of short skip conditions on 40m leading to only 104 contacts, including only 14 mill stations.

On VHF, activity was similar to previous years. Using the Society's FT-857 transceiver and 6m-2m-70cm vertical, a total of 65 stations were contacted with the bulk of traffic on 2m FM, but a 6m Sporadic-E opening provided some more distant contacts into Eastern Europe.

The event was again popular, with a steady stream of visitors keeping the team busy on the Sunday.

Rejuvenated DF Hunt

After some poor support for annual DF hunts, and the committee's decision not to organise an event in 2015, Guy G0UKN devised a refreshing new format for 2016, which saw seven teams successfully locating two low power 2m FM transmitters which were sending out a Morse ident every two minutes.

On a fine summer evening, 20 members gathered next to the *Crown Inn* on Chislehurst Common. Unlike previous hunts, teams could leave when they were ready, given a start time when they set off and a finish time when they returned. The clear winner of 'The Tally Ho! Cup' was Colin G3SPJ, who used an FT-817 and

small homemade loop antenna to locate the two transmitters. Several of the teams had never taken part in a DF hunt before, so the new format was judged a success.

Colin G3SPJ receiving 'The Tally Ho! Cup' from event organiser, Guy G0UKN

NFD from the valley of the River Cray

Society Chairman, Dave G4BUO, was keen for participation in all the three field days in the 70th anniversary year. However, it was not possible to enter SSB Field Day due to lack of support, but there was a successful return to NFD. It was fitting that in the Society's 70th year, the G3RCV/P activity was from the valley of the River Cray in Bexley. The site, which had been found by Lawrie G4FAA, was close to a well-stocked farm shop with an on-site restaurant which served Sunday roast! Research after the contest found that the farmer's grandfather was A.K Wall, G2YZ, a member of the Society from 1946 to 1957 and a committee member in 1951/52.

Mark M0DXR and Dave G4BUO used a 270ft doublet antenna and 100 watts from a Kenwood TS-990 transceiver. Conditions were not good, but they made 1,154 contacts in the 24-hour contest period. This secured overall sixth place in the

Restricted section, but third in the 'simple antenna' section. G3RCV/P also received a certificate for being the 80m band leader in the 'simple' category.

A satisfying VHF NFD

Lawrie G4FAA also secured the use of the Kent County Showground at Detling for the Society's entry into VHF NFD. With enough interest and equipment, Colin G3SPJ organised a four band entry in the restricted section.

Not long into the contest, it became apparent that VHF conditions were poor, with low activity. There was no Sporadic-E propagation either. That was also the case for the 4m part of the contest on the Sunday, using 40w from a new Icom IC-7300 into a 7-element yagi. In the circumstances, a good team effort saw G3RCV/P placed 5th.

Steve M0SCJ operating at VHF NFD

South East Tutors training group formed

The Society hosted a meeting of local amateur radio societies – Cray Valley, Crystal Palace, Darenth Valley, North Kent and West Kent to agree the potential benefits of working closer together as a group (christened the South East Tutors) to plan and deliver club based training. An agreed timetable for courses into 2017 was agreed.

70th anniversary preparations

Detailed planning for the October celebrations was well under way by July. A special callsign – GB70CV – had been granted by Ofcom; the rules of the 'Platinum' award and associated activity periods had been agreed; and special anniversary goods with a special 70th anniversary logo were available for members to order. Invitations to the special celebratory 70th anniversary party had been issued to 170 current members, ex-members and guests; the President of the Radio Society of Great Britain, Nick Henwood, and Eltham Member of Parliament, Clive Efford, had been confirmed as guests of honour.

As the party approached, over 100 acceptances had been received; Trevor M1TAD had arranged operating rotas for GB70CV and the Society callsigns G(X)1RCV and G(X)3RCV; a celebratory cake had been ordered from *J Ayres* in Eltham (bakers who had baked the Society's 60th anniversary cake); over 80 items of anniversary goods had been ordered by members; Chris G0FDZ's USB archive contained over 60GB of Cray Valley memorabilia; and this book had been largely completed!

The 'Platinum' anniversary month

The month began with GB70CV active from the Society's shack at the 1st Royal Eltham Scout Hall. In 13 hours of operation on 1 October Nobby G0VJG, Richard G7GLW and I made 1,000 contacts. A number of members operated GB70CV on many occasions through the month, making close to 3,000 contacts. Amateur radio was demonstrated to local community groups, Scouts, the local Member of Parliament and the local press. The Society callsigns, G(X)1RCV and G(X)3RCV were also active during the month, making a total of over 4,000 contacts from various members' homes.

Wartime radio talk: The first meeting of the anniversary month saw a talk by Smudge G3GJW. It provided the best attendance for many years. Over 40 members attended the meeting and heard Smudge's reminiscences of wartime radio, the equipment in use during the 1940s and 1950s, and some insight into the early days of the Society.

Activity periods: A series of 11 activity periods were arranged for the following week, to make it a little easier to qualify for the 'Platinum' award, a specially engraved shield awarded for achieving 70 points by working GB70CV, G(X)3RCV, G(X)1RCV and 10 member stations. Activity was on 80, 40, 20, 10 and 2m. The 40m and 2m events were the best supported, with up to 20 member stations active during several of the periods.

Due to their popularity, and because some members had not obtained sufficient points, three further periods of activity were arranged for later in the month. At the time of going to press, 25 'Platinum' awards claims had been received.

The 59th JOTA: GB2RE was activated on behalf of the Severndroog Explorer Scout Unit from Severndroog Castle on Shooters Hill. 100 Beavers, Cubs, Scouts and leaders learnt about Morse, took part in a DF hunt and passed greetings messages by amateur radio to other Scouts taking part in the event.

Guy G0UKN with a party of Scouts at GB2RE

70th anniversary party: 100 members, ex-members and guests attended the Society's 70th anniversary party. It was a real pleasure to welcome a number of ex-members from the earlier days of the Society. In particular, it was good to greet Richard G3TFX, Ian G8CPJ, Glyn ex-G8KKI, Peter G3RZP, Kevin G8KDC, Don G4KOW, Terry G3VFO, Norman G3ZCV and Lynne G4FNC.

Some of the guests at the anniversary party

A popular attraction through the evening was the display of photographs and memorabilia I had compiled of the Society's history through its 70 years. GB70CV was also on the air during the evening, and a number of members including Richard G7GLW, Mick M0XBF and the Society's newest member, Andrew G4ADM, enjoyed making contacts.

The anniversary display charting 70 years of Cray Valley history

Chris G0FDZ was MC for the evening. After his opening address, he introduced Dave G4BUO (Chairman), Nick Henwood G3RWF (RSGB President), Clive Efford (MP for Eltham) and myself (President). After four short addresses, those present enjoyed ample time mingling with others, while enjoying the refreshments organised by Nigel G1BUO, with help from Dave M6FDG and Frank G0FDP.

Clive Efford sharing a witty moment with the audience

L-R: me, Chris G0FDZ, Clive, Nick G3RWF and Dave G4BUO

A donated raffle, organised by Giles M0TGV and Richard G7GLW, raised over £500 for Society funds. My daughter, Clare, realised over £200 in selling anniversary goods, taking orders for this history of the Society, and accepting a number of 'Platinum' award clams.

As the evening drew to a close, Nick Henwood G3RWF proposed a toast to the Society and Clive Efford cut the celebratory cake.

The 70th anniversary cake

...and so the lights went out on a truly memorable evening that achieved widespread acclaim from those present. It was a wonderful occasion that passed all too quickly, but Cray Valley's special anniversary had been marked in true high-profile fashion.

CQ Worldwide SSB contest

The 70th anniversary month ended with an entry into the SSB section of the CQ Worldwide DX contest. Tim G4DBL aired the Society's contest callsign M8C from a location in the Chiltern Hills.

Using a log-periodic antenna at 40 feet for the HF bands and a 264 foot dipole at 60 feet for the LF bands, he enjoyed some rapid-fire pile-ups when the 20m trans-Atlantic path opened. He also had plenty of first call responses on 10 and 15m when reverting to 'Search & Pounce', but he found LF conditions disappointing. Indeed, band conditions were generally poor during the entire weekend, with no sunspots at all on day one.

It was not a winning entry, but Tim was pleased to have had the opportunity to use the contest callsign for the first time in its anniversary year.

Reflection and acknowledgements

All that remains is to wish the Society
continued success for the future. It was a
pleasure researching its history to be able to
compile such a detailed account of Cray
Valley Radio Society and its activities
through its 70 years.

In closing, I acknowledge in particular the
help and encouragement received from
Chris G0FDZ, Dave G4BUO, Smudge G3GJW,
Mark Allgar at the RSGB, my daughter Clare,
and all those who in some way have
contributed to this significant history of,
arguably, one of the UK's major amateur
radio societies.

*The author wishing Cray Valley Radio Society
success for the future*

Cray Valley Radio Society Presidents and Chairmen

Presidents

1965-1968	Arthur Milne G2MI
1969-1977	Stan Coursey G3JJC
1978-1993	Alan Swindon G3ANK
1993-	Bob Treacher BRS32525/M3RCV/2E0RCV/M0MCV

Chairmen (elected at April Annual General Meetings)

1946-1949	Ernie Redpath G2DS
1949-1951	No records available
1951-1953	Ernie Redpath G2DS
1953-1962	No records available
1962-1965	Bill Green G3FBA
1965-1966	Stan Coursey G3JJC
1966-1967	Brad Bradwin G3DNC
1967-1968	Stan Coursey G3JJC
1968-1969	Ken Wooff G3TCC
1969-1971	Fred Tickner G3XFG
1971-1972	Deryck Buckley G3VLX
1972-1974	Derek Baker G3XMD
1974-1975	Martin Tripp G3YWO
1975-1978	Bob Treacher BRS32525
1978-1981	Bernard Harrad G8LDV
1981-1982	Owen Cross G4DFI
1982-1983	Alan Burchmore G4BWV

/cont.

1983-1985	Phil Wolfe G4EGU
1985-1986	Graeme Caselton G6CSY
1986-1987	Chris Whitmarsh G0FDZ
1987-1988	Brian Rowe G4WYG
1988-1990	Chris Whitmarsh G0FDZ
1990-2000	Nigel Cornwell G1BUO
2000-2010	Bernard Harrad G8LDV/G3NPS
2010-2012	Kevin Jennings M0KSJ
2012-2013	Guy Roberts G0UKN
2013-2015	Frank Rhodes G4WNF
2015-	Dave Lawley G4BUO

Cray Valley Radio Society: Membership 1946 - 2016